The Agenda in Action

The Agenda in Action

1983 Yearbook

Gwen Shufelt
1983 Yearbook Editor
University of Missouri—Kansas City

James R. Smart
General Yearbook Editor
San Jose State University

National Council of
Teachers of Mathematics

Library of Congress Cataloging in Publication Data:

National Council of Teachers of Mathematics.
The agenda in action.
(Yearbook ; 1983)
Bibliography: p.
1. National Council of Teachers of Mathematics.
An agenda for action : recommendations for school mathematics of the 1980s. 2. Mathematics—Study and teaching—United States. I. Shufelt, Gwen. II. Series: Yearbook (National Council of Teachers of Mathematics) ; 1983.
QA1.N3 1983 [QA13] 510s [510'.7'1073] 82-22367
ISBN 0-87353-201-5

Printed in the United States of America

Contents

Recommendation 2

Recommendation 3

Recommendation 4

Recommendation 5

Recommendation 6

Recommendation 7

Recommendation 8

Preface

The Agenda in Action is a particularly appropriate topic for the NCTM's 1983 Yearbook for these reasons:

- Interest in *An Agenda for Action: Recommendations for School Mathematics of the 1980s* will continue to grow, since the recommendations focus on a decade and perhaps beyond, not just one year.
- The *Agenda* recommendations, to be effective, must be implemented by many teachers and other school personnel in many individual classrooms, schools, and communities.
- It is time for a progress report on the *Agenda* in action, one that stresses good ideas that have worked for others and that you might want to try for yourself.
- The *Agenda* recommendations provide a convenient framework for exploring current developments leading to the improved teaching of mathematics at all levels.

Because of the timeliness and appropriateness of this topic, the Educational Materials Committee of the NCTM has selected "the *Agenda* in action" as the theme of the 1983 Yearbook. The variety of backgrounds that the authors of these articles bring to this book is indicative of the number of individuals who are currently using the *Agenda* to improve mathematical instruction. We hope the content in this yearbook will serve to increase the rate of realistic implementation of the NCTM recommendations.

After the Introduction, which describes many facets of current action on the *Agenda,* all other articles in this yearbook are presented under eight headings—the eight recommendations in the *Agenda.* Although some of the articles are related to more than one of the recommendations, each is placed in the section for the major idea addressed. Section 1 has a general article on problem solving followed by six articles that give specific suggestions on teaching problem solving at every level from early childhood and primary grades through teacher education classes. Section 2 includes a variety of articles illustrating the NCTM position that teaching basics means much more than practicing numerical computation. Section 3 shows some current ideas for incorporating the technology of calculators and computers into mathematics instruction.

Section 4 shows examples of instructional strategies and activities that promote an effective learning environment. Section 5 provides encouraging new approaches to evaluating mathematics students and also gives insights into how to interpret test results. Section 6 reports promising attempts to implement a more flexible curriculum in mathematics. Section 7 looks ahead to what the next decade may hold in store for the teaching profession. Section 8 takes into account how much we need the aid of additional people and groups outside the NCTM if the *Agenda* is to have a long-lasting, positive effect.

The lengthy but rewarding task of reviewing the many proposals, selecting the completed articles to be included, and making suggestions for revision of those selected was carried out by the editors and a carefully chosen review committee. Our thanks and appreciation go to the review committee members:

Hilde Howden, Albuquerque (N. Mex.) Public Schools

Helen Kriegsman, Pittsburg (Kans.) State University

Earl Ockenga, University of Northern Iowa

William C. Smith, University of Missouri—Kansas City

The editors also wish to thank the NCTM Educational Materials Committee and the NCTM headquarters staff for their support and guidance through the process of developing the book. We further want to thank each author represented here and all others who submitted proposals or manuscripts for consideration.

We are pleased to present this completed yearbook for your use. We believe it will serve as a reference and a source of new mathematical ideas and activities over a period of many years. We sincerely hope that this material will make possible a great range of significant improvements in the teaching of mathematics.

GWEN SHUFELT
1983 Yearbook Editor

JAMES R. SMART
General Yearbook Editor

Introduction

An Agenda for Action: Status and Impact

Shirley Hill

RECOMMENDATIONS and position papers do not arise in a vacuum. *An Agenda for Action: Recommendations for School Mathematics of the 1980s,* like all such documents, was a response to particular demands and pressing concerns. Throughout the decade of the seventies, the mathematics education community seemed to be groping for a clearer focus and sense of direction. Its perception of this need was evidenced by many conferences, by individual articles, and by the report of the National Advisory Committee on Mathematics Education, Conference Board of the Mathematical Sciences, *Overview and Analysis of School Mathematics, K–12.*

The National Council of Teachers of Mathematics (NCTM) has a respected history of responsible position papers, guidelines, and recommendations. The *Agenda for Action* and its recommendations fit perfectly into the mold of professional leadership. The document is at once an official position of the NCTM and a guide for concerted action. As such, it is not just an agenda for the organization's activities but the suggested agenda for a decade's effort by all persons and groups concerned about mathematics education. It not only serves as a focus for NCTM activities but also sets some goals that can only be realized by all sectors, public and private, in cooperation. Of course the *Agenda* is ambitious, but it represents aggressive leadership.

One can argue that the preeminent role of a professional organization such as the National Council of Teachers of Mathematics is that of leadership. This need not downplay those activities more visible to the membership in which ideas and materials are shared with other teachers through journals, other publications, and meetings. This continuing "in-service education" function makes a significant contribution to the NCTM goal of improving mathematics education by upgrading individual teaching skills. But the classroom is not an isolated unit impervious to the influences, pressures, and even whims of broader policy, trends, and public opinion. To be serious about improving mathematics teaching and learning on a broad scale, mathematics teachers and educators need to take—and present collectively—positions designed to influence policy decisions. The obligation of a professional organization is to offer its considered advice in its area of expertise to decision makers and to the public and the public's representatives.

The NCTM was not the only organization to respond to the perceived need for a clearly delineated direction. The National Council of Supervisors of Mathematics developed a *Position Paper on the Basics in Mathematics,* which was an influential answer to the helplessness many mathematics educators felt in the face of what appeared to be extreme public pressure to narrow the definition of the "basics" in school mathematics.

Concern for direction for the 1980s was reflected in a conference sponsored by the Mathematical Association of America, Priorities in Mathematics Education in the 1980s (PRIME 80). This conference resulted in a number of recommendations that bear on high school preparation for college mathematics.

These conferences and papers date from the late 1970s. Earlier, at mid-decade, NCTM's Board of Directors had adopted as its primary five-year goal the development of a set of curriculum recommendations for the 1980s. The target date was set at 1980. The goal was later expanded to encompass a broader set of policy recommendations, not confined to the narrow definition of curriculum. This change was significant in that it signaled the Council's determination to address itself to policy makers and the wider audience of those who influence educational priorities—not just to those within the profession itself.

Recognizing the need for a broad range of input from a cross section of professionals and laypersons, the Council collaborated with the Ohio State University in a project to assess priorities and preferences in mathematics curriculum and instruction. The project, Priorities in School Mathematics (PRISM), was a major contributor to the data base studied in developing the *Agenda for Action* and continues to be a source of valuable information in planning strategies for implementation. The results are especially useful in identifying levels of support for particular recommendations and pinpointing groups in which the strongest support lies. Similarly, groups in which support needs to be developed are identified.

The organizational apparatus for developing the specific recommendations consisted of three ad hoc committees, with the president of NCTM as liaison among the three committees and between these three committees and the standing committees of the Council.

Overseeing the operation was a Board of Directors subcommittee, the Mathematics Curriculum for the 1980s Committee. Most of the members of this group were also on the steering committee for the PRISM project. This assured direct communication of PRISM planning and results to the recommendation development project. The Board of Directors also appointed a Task Force on Recommendations. Its charge was to gather and study all pertinent information and data and to prepare the *Agenda* document and recommendations. This group decided on format and categories of recommendations and did most of the writing, with considerable interaction with the Mathematics Curriculum for the 1980s Committee. The Curriculum Committee participated significantly in the process of development, both in the early planning stages and in suggestions and some revisions of the several drafts. The Board of Directors reacted to preliminary drafts with suggestions and approved the document, making it the official NCTM position.

It was recognized early that the *Agenda,* in order to exert the influence desired, must go beyond one Board's official imprimatur. It needed to reflect, as nearly as possible, a collective position of the membership. It could not represent only the thinking of a few committees. A true concensus can never be claimed for such a large organization, but it is possible to discern a general "sense of the membership" by tapping the thinking of representative groups.

Members of the Task Force and Mathematics Curriculum for the 1980s Committee attended many NCTM meetings and talked with many people. The president's role as liaison between the ad hoc committee members and standing committees was useful. During the process of development the standing committees were, of course, carrying out their regular business, and their meetings with the NCTM president in attendance provided the opportunity for input, feedback, and the testing of ideas.

Additionally, a questionnaire was sent jointly by the chairman of the Mathematics Curriculum for the 1980s Committee and the NCTM president to all presidents of affiliated groups urging a response that reflected their leadership and, where possible, their membership. Over three hundred questionnaires were returned, many representing a consensus or summary of the thinking of a large group of members.

Finally, the president's report in one article of NCTM's *Newsletter* urged individual members to write and express their recommendations. Few people took advantage of this opportunity, perhaps because the request was not specific enough. Nevertheless, the NCTM membership's response to the *Agenda* has been, in general, very favorable, which may be one piece of evidence for my belief that the *Agenda* did capture in its major themes the mainstream of Council opinion.

Contact with sister organizations in mathematics and education was maintained by having an officer of the Mathematical Association of America serve as a Task Force member and the NCTM president sustain active membership in the coalition of educational organizations called Organizations for the Essentials of Education.

Although the *Agenda for Action* is not in the strictest sense an application of research, the committees involved directly in its development sifted through, analyzed, and were directed by a substantial collection of recent information. Most prominent in the data base, in addition to the PRISM data, were the results of the second mathematics assessment of the National Assessment of Educational Progress, three status studies funded by the National Science Foundation, and preliminary results of John Goodlad's massive study of what goes on in the classrooms of the United States.

Implementation and Prospects

The *Agenda for Action* makes clear that its purposes were not served merely by its publication. The key first step, and a necessary condition to the achievement of those purposes, was distributing it widely and selectively. The initial printing was for 100 000, and all NCTM members were mailed copies. Copies were also made available at the Annual Meeting in 1980.

This distribution was an essential beginning—but only a beginning. The NCTM membership comprises a minority of secondary school mathematics teachers and a tiny fraction of elementary school teachers. But here we need to be reminded of the *Agenda*'s intended function. It is a function of leadership, and the members of organizations like NCTM are the professional leaders among teachers. The intent of the *Agenda* is that we convey a message *from* teachers, not *to* teachers. It is the leadership of our profession speaking outside itself to leaders in the decision-making process in education and to the concerned and *attentive* public. Thus distribution needs to be, not exclusive, but carefully selective.

The *Agenda* was distributed to educational administrators, school board members, PTA leaders, government officials at all levels, and publishers. Thousands of copies have been ordered and disseminated by school districts, publishers, and state and local mathematics organizations.

Implementation efforts by the NCTM fall roughly into five categories. This state-of-the-*Agenda* report is organized accordingly.

1. *Public relations.* Part of the activity in this category is the wide distribution already described. Another goal is to assure coverage and references in many publications. A press release was issued by NCTM at the time of the *Agenda*'s initial presentation. Articles appeared in many newspapers including those with national circulation, such as the *Christian Science Monitor.* There has been coverage in specialized popular magazines, such as *Science News* and *Education USA,* and in professionally oriented journals, such as the *Curriculum Report* of the National Association of Secondary School Principals. Editorial comment has appeared in publications of many professional organizations, such as *Basic Education* of the Council for Basic Education and the *SIAM News,* the newspaper of the

Society for Industrial and Applied Mathematics. Editorial comment has been, in general, very favorable.

Presentations have been made to a wide variety of organizations, among them the National School Boards Association, the Association for Supervision and Curriculum Development, the National Association of Independent Schools, and most school administrators' organizations. Several conferences have been organized around presentations and discussions of the recommendations, notably the Wesleyan University Conference, which brought in executives from high technology industry, and the Conference of the Virginia Council of Teachers of Mathematics, which included a large number of state and local educational leaders.

2. *Political action.* At the level of the federal government, the *Agenda* has been delivered to all congressional offices, either to the senator or representative personally or to a member of his or her staff. Staff members in the executive branch have also received copies. Organizers of a conference of the National Institute of Education explicitly asked presenters to consider one or two of the *Agenda*'s recommendations and distributed copies to participants.

At the state level, many official documents have cited the *Agenda for Action.* A publication of the Florida Department of Education endorsed the *Agenda* and reprinted it in its entirety as an appendix. The curriculum guides of the Texas Education Agency also include an endorsement of the *Agenda* and print the eight major recommendations as part of the guides. One cannot be certain that the Council's recommendation on increasing high school mathematics to three years had a direct influence on California Governor Jerry Brown's assertion that one goal for education should be the requirement of three years of high school mathematics for all students, but it is known that the members of the California Mathematics Council made sure his staff had copies of the *Agenda.* And reference to the *Agenda* was made in the report by the Ohio Commission on Articulation between Secondary Education and Ohio Colleges, a report that recommends three years of high school mathematics for all college-bound students.

Another aspect of political action is that of gathering the support of other groups. Formal endorsement of the *Agenda* came from the Conference Board of the Mathematical Sciences, an umbrella group of all mathematics organizations, and the International Reading Association. The leaders of many other organizations have noted its publication favorably.

3. *Support for local efforts.* This category recognizes an important function of position papers like the *Agenda for Action.* It is not just that they attempt to present new ideas or concepts but that they convey a professional consensus that by its very statement can provide support for those who are trying to accomplish things educationally in their own locales. Obviously, those re-

sponsible for preparing the recommendations did not discover or originate all the ideas contained therein. But leadership consists in recognizing a surge of ideas— a gathering momentum, a series of trends that coalesce around a general principle—and pulling these things together into a statement, a statement that represents a profession.

There are more instances than anyone can collect of local efforts that do not owe their genesis strictly to the NCTM *Agenda for Action* but that have found the recommendations a useful support for their activities. The NCTM Mathematics Education Trust is supporting two such efforts: one, the development in Wisconsin of a three-year curriculum of general mathematics, and another, a project in Virginia to prepare a monograph on the *Evaluation of Mathematics Learning.* The monograph responds to the recommendation that "learning be evaluated by a wider range of measures than conventional testing."

4. *Collection and dissemination of model programs.* The commissioning of this yearbook by the Board of Directors is a case in point. The eventual publication of the results of the two projects mentioned above also fits into this category of implementation. There are a number of publications in the NCTM pipeline that provide models.

It is early yet to see many models of implementation. As we move on toward mid-decade, it will be incumbent on the NCTM to publicize and encourage a sharing of the promising ones.

5. *Production of guidelines and instructional resources.* This category fits directly into the ongoing activities of the NCTM, not just in publications, but in committee and task force reports, conferences, and seminars. If one scans NCTM meeting programs, publication lists, and news bulletins, the extent of the effort and dedication to the implementation of the *Agenda for Action* recommendations is evident.

Two immediate responses were the publication of the guidelines for selecting computer software and the professional reference series book *Changing School Mathematics,* copublished with the American Association of School Administrators and the Association for Supervision and Curriculum Development. The latter is a textbook, indeed a must, for anyone who wants to accomplish change in the mathematics curriculum. And that, of course, is what the *Agenda for Action* is all about.

State of the *Agenda,* 1983

We are one-third of the way through the decade of the eighties. Are we on schedule? What have we accomplished? The fact is that we can never be sure. Education is far too complex to trace direct causation for broad movements and ambitious objectives. Things change, but many, many factors contribute. As it is with our teaching, we never know the direct influence or product of our efforts.

A wag once said that being a leader means determining which way the mob is running and then running out ahead. In some ways, professional leadership must do some of that. Perhaps our recommendation on using computers has a little of that flavor. Are we just riding the crest of a wave or are we helping direct it? Probably it is a little of both. To lead, we sense the coalescing of professional opinion and then help guide the movement responsibly.

There is no question that references to the *Agenda* abound; one sees them everywhere. There is no question that we see evidence—in textbooks, in school programs, in statements from leaders—that things are happening and in the direction of the NCTM recommendations. Are they happening too slowly? Change in education always seems to move ponderously. One-third of the way through our decade, I think we are right on schedule and certainly still on task.

You can be sure there is no loss of commitment. In 1982 the NCTM Board of Directors created a select committee to formulate an "action plan" for continuing the implementation of the *Agenda.* Its responsibilities to oversee the activities and advise the Board accordingly will continue throughout the decade.

Recommendation 1

PROBLEM SOLVING MUST BE THE FOCUS OF SCHOOL MATHEMATICS IN THE 1980s

1

Problem Solving as a Focus: How? When? Whose Responsibility?

Peggy A. House
Martha L. Wallace
Mary A. Johnson

TO MAKE problem solving the focus of school mathematics . . . to apply stringent standards of efficiency and effectiveness to the teaching of mathematics . . . to demand of ourselves and our colleagues a high level of professionalism—these are laudable and challenging goals. But how shall we translate them into actual classroom practice that will change the behaviors of both teachers and pupils? Therein lies the real challenge of the *Agenda for Action*. To investigate this challenge, we must first understand the complexity of teaching in the style implied by the *Agenda* and then discover ways to implement the recommendations. Let us begin this investigation by reflecting on the experience of a student teacher.

Clarifying the Problem

John was perplexed as he shared the experience of his first day of student teaching. He had been observing a junior high school mathematics class when he was approached by Del, an eighth grader, who began to ask questions about courses in high school and college mathematics. "I'm going to be an architect," Del announced, "so I'll need lots of math, won't I? I think I need trigonometry and calculus. Do you know what they are?"

As John was struggling to find a way of explaining trigonometry to an eighth grader, Del was already asking, "Can you show me some calculus problems?" John showed Del how to find the area under a curve by adding rectangles, and Del soon concluded that one could probably use a similar technique to find the volume of a mountain. As the class ended and Del was leaving, he turned to thank John for helping him and to share his philosophy of what makes a good teacher: "Teachers spend lots of time telling us things we should know. I would like to have teachers who want to make it stick!"

"I was really excited," John recalled later. "My first day out, and I had found a student who was excited about mathematics, a student who was a good problem solver! But later when I told Del's teacher about my discovery, she said, 'Del's no problem solver! He can't even add fractions yet, and he needs to know how to do that before he can solve problems!' "

Who was right, John or the teacher? Is it true that Del was not a problem solver even though he seemed much more interested and excited about mathematics than most of the students one observes? To answer these questions, we must define "a good problem solver" and decide whether we can identify a successful problem solver in such a short encounter. We must also ask, "What is problem solving?"

The common definition of a mathematical problem is a situation that involves a goal to be achieved, has obstacles to reaching that goal, and requires deliberation, since no known algorithm is available to solve it. The situation is usually quantitative or requires mathematical techniques for its solution, and it must be accepted as a problem by someone before it can be called a problem. Problem solving is the process of attacking such a problem: accepting the challenge, formulating the questions, clarifying the goal, defining and executing the plan of action, and evaluating the solution. It will involve the use of heuristics but not in a predictable manner, for if the heuristics could be prescribed in advance, they would themselves become algorithms and the problem would become an exercise.

Problem solving is a process, not a step-by-step procedure or an answer to be found; it is a journey, not a destination. Successful problem solvers can be identified by the processes or the attitudes of mind they display. Four characteristics that help identify good problem solvers are desire, enthusiasm, facility, and ability.

Recognizing a Problem Solver

A problem solver needs, first of all, the *desire* to approach the problem, accept a challenge, take a risk, find an answer, understand a question, and discover new knowledge or create a new solution. A problem does not exist until someone accepts it as a problem, and a problem solver does not exist without the desire to solve problems.

Along with this desire, a problem solver needs *enthusiasm* to proceed with the solution. Enthusiasm, as we use it here, signifies the willingness to accept a challenge or set one's own challenge. It means having the determination to investigate past the first obstacle and the perseverance to continue even when the problem looks hopeless. It means possessing the flexibility to try several methods and to look for more questions once the original problem has been solved. Enthusiasm is kindled by successful problem-solving experiences, so that the problem solver knows the delight of conquering a problem.

Problem solving also requires *facility* in using mathematics and heuristics. Facility with mathematics includes understanding fundamental concepts, relationships, and mathematical processes; it means "basic skills" broadly defined. Facility with heuristics signifies the ability to make guesses; solve simpler or related problems; construct pictures, graphs, tables, and charts; recognize and generalize patterns; make and test predictions; offer explanations; and apply results to new situations.

A successful problem solver sees the general structure of problems, separates meaningful data from irrelevant detail, thinks of several ways to approach the problem, and extends the problem and its solution into other areas. Thus, the problem solver must have a variety of techniques with which to attack a problem and the understanding needed to judge which techiques are likely to work and which will probably not. And since a problem is not solved until the mathematical aspects of the situation are worked through, the problem solver must not only know appropriate mathematical concepts but also be able to evaluate the acceptability of the solutions.

Finally, problem solving requires *ability*. Just as there seem to be naturally talented artists, musicians, and athletes, so there seem to be naturally talented problem solvers; these outstanding problem solvers, like the outstanding artists, musicians, or athletes, are sometimes thought to be born, not made. This may be true of a brilliant few in each field, but all people possess some measure of ability and this ability frequently goes undeveloped. Just as ordinary persons can learn to enjoy painting, piano playing, or tennis, so ordinary students can learn to enjoy mathematics. Just as many less-than-outstanding persons can, with training, become good artists, musicians, or athletes, so, too, can students become good problem solvers when they are given the opportunity to develop whatever talents they have. Likewise, even outstanding problem solvers can become better. Problem solving is not the domain of a select few but the right of all mathematics students.

Then what of Del? Is he a good problem solver by our definition? Probably not. True, he exhibited some measure of desire and enthusiasm; yet, according to his teacher, he still lacked certain kinds of facility, and we have very little indication of his ability. It would seem that Del is not yet a problem solver. Can teachers make Del and others like him into better problem solvers? How? And most important, how will teachers develop the skills they need to achieve the goal of teaching with a problem-solving focus?

"Making It Stick"

Del's characterization of a good teacher was one who could "make it stick." How does a teacher go about meeting Del's challenge? How, indeed, do mathematics teachers develop problem solvers with desire, enthusiasm, facility, and ability?

First, teachers must foster the desire to approach, accept, and try to solve problems. They must stimulate the curiosity that is in every student and direct that curiosity toward mathematical problems. Second, teachers must model enthusiasm for problem solving, showing students how determination and perseverance in attacking a problem often lead to the delight of solving it. Third, teachers must help students develop the facility in mathematical and heuristic skills that will enable them to solve new problems. Unlike Del's teacher, they should view a lack of facility not as a deterrent to problem solving but as an opportunity for encouraging new learning in order to make problem solving more possible. Finally, teachers must nurture their students' ability for problem solving by helping them find and develop their own strengths, recognizing their differences, and finding ways to challenge each class member in ways commensurate with the individual's ability and interests.

And how are these outcomes to be realized? Not by waiting for extensive curriculum changes, nor by adding puzzles and routine story problems to lessons, nor even by teaching heuristics to students. Some curriculum changes may be desirable, but problem solving can become the focus of mathematics without extensive curricular reform. Puzzles are fun and can be used for diversion, motivation, and enrichment, and story problems can be useful for practice and the reinforcement of procedures; yet neither is enough to make students better problem solvers. Heuristics are necessary to successful problem solving, but if they are isolated topics of instruction, then problems become exercises and the heuristics act as nongeneralizable algorithms.

In assessing the modern mathematics movement of the 1950s and 1960s, E. G. Begle once observed that mathematics educators had made progress toward solving the problem of teaching better mathematics but not of teaching mathematics better. The primary challenge of the 1980s continues to be to learn to teach mathematics better. Achieving the *Agenda* goal of making problem solving the focus of mathematics instruction demands significant changes in teacher behavior. The curriculum can be left largely unaltered, and existing textbooks can still be used (although, no doubt, used differently). But teachers cannot persist in employing the commonly observed repetitive routine of correcting the homework, going over a few examples, and assigning more homework; nor can they continue to represent mathematics as essentially computational, consisting of precise rules and algorithms and primarily justified by the importance of each topic for subsequent topics or courses.

Teaching with a Problem-Solving Focus

Implementing the problem-solving goal of the *Agenda* depends on the instructional strategies employed by each individual teacher. Consider Jane, a successful teacher with a problem-solving focus. At first glance, Jane's classroom looks like many other first-year algebra rooms. Her bulletin boards are colorful; one contains a section called "Problem of the Day" (fig. 1.1) and another called "Problem of the Week" (fig. 1.2). Jane uses a popular current textbook. Most days she requires homework from the students, and she is planning to include questions on the unit test requiring familiarity with common algorithms.

Problem of the Day

Mark and Margie ran a 100-meter race. Margie crossed the finish line while Mark was still at the 95-meter point, and so she won the race.

In a second race, Margie gave Mark a 5-meter handicap by starting 5 meters behind the START line. Both ran at the same constant speeds that they ran in the first race. Who won the second race?

Fig. 1.1

A closer look reveals Jane's emphasis on problem solving. The bulletin boards contain not only "trick puzzles" but real problems, frequently with extraneous information or with unstated questions. Two problems are presented together, and the students are asked to find similarities and differences in the problems themselves rather than in the answers. Sometimes solutions are posted, but the challenge is to find another method of solution or to create another problem with a similar solution. Problems seem to reflect student interests, and student-produced problems are often posted. Class time is set aside to discuss solutions, solution methods, and the reasons for selecting those methods. Students are encouraged to generalize results and to investigate similar problems each time a problem is solved.

Jane uses the textbook in her teaching, but she often starts at the middle or end of a chapter and works backward. She began the present unit on linear equations by sending the students home with a variety of work problems from

Problem of the Week

The five dart boards shown here are all squares with perimeter 48 units. In each board, the circles are congruent and tangent to each other and to the edges of the boards.

Darts are thrown without aiming, and so each dart has an equal chance of hitting anywhere on the board.

1. If a game scores one point for each dart that hits inside a circle, which board would you rather play on?
2. If you throw 100 darts, how many points do you expect to score?
3. If a game consists of 100 points, how many darts would you expect to throw?
4. If you get +1 point for each dart inside a circle and −2 points for each dart outside a circle, what score do you expect after 200 throws? How many throws do you expect to need to score 100 points?
5. If a dart hits the dot in the center of a circle, you score +100 points, but if it hits any point on the circumference of a circle, you score −10 points. Under these conditions, which board would you choose to play on?
6. In a game for three players, scoring is as follows:

   Player A: +1 for each dart inside a circle
   0 for each dart outside a circle

   Player B: +3 for each dart outside a circle
   0 for each dart inside a circle

   Player C: +2 for each dart inside a circle
   −4 for each dart outside a circle

   Which player do you expect to be the winner?
   Does your answer depend on which dart board you choose?
7. Make up a set of scoring rules for a fair game for two persons.

A

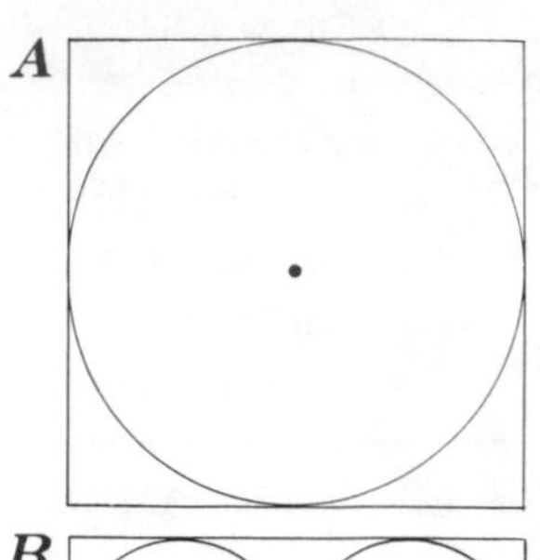

B

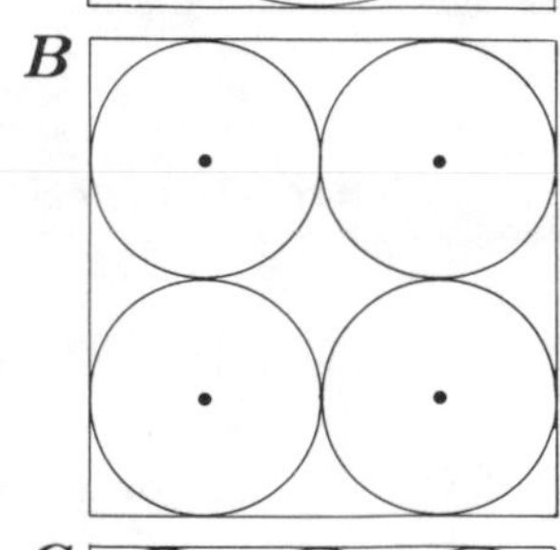

C

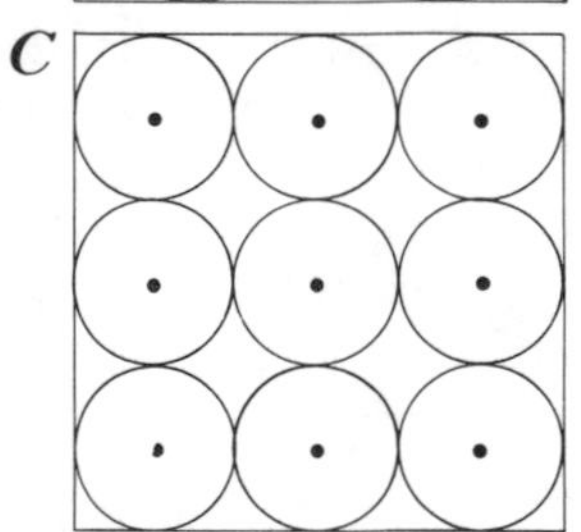

D

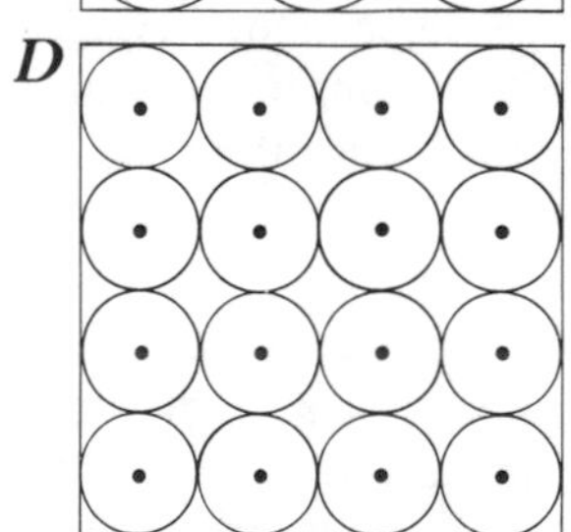

E

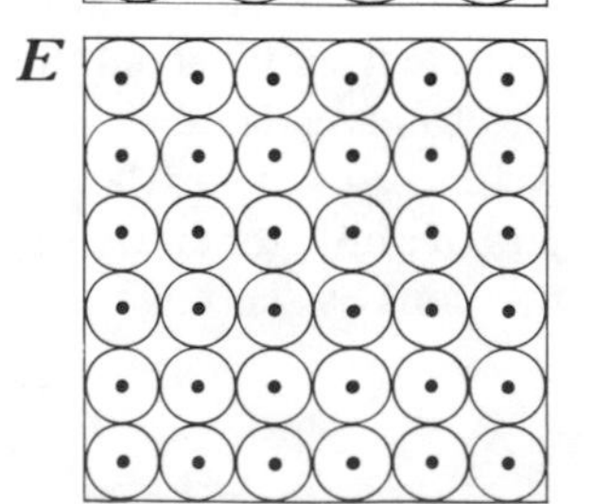

Fig. 1.2

the end of the chapter. Since the students did not yet know the applicable algorithms, these were truly problems, not exercises as they would be later. Students were told to work as many of the problems as they were able, to classify the others, and to tell what they needed to know to solve the problems in each category. Thus they received practice in identifying common structures and in recognizing what they needed to know as well as receiving motivation for studying problems involving linear equations.

Today the class is conducting experiments to measure the stretch of various springs as different weights are added. The students are trying to predict results for cases they have not measured, a task involving the solution of linear equations even though they have not been given the algorithms. Later the students will play a variant of "Guess My Rule" where they already know the rule but must guess the input, given the output.

As the students become more adept at solving the equations generated through experiments, games, and textbook exercises, Jane will focus their attention on the methods they are using. Discussion will center on the different ways students are finding answers, and the class will investigate which methods are most efficient, most reliable, or easiest to use and remember. Only after this discussion will the students be told to work through the early part of the chapter where they will see the author's development of the algorithms and have a chance to practice their own.

One gets the impression while watching Jane teach that she genuinely enjoys mathematics and problem solving. She always seems excited about whatever problems her students are solving, and she usually has one of her own to work on in her spare time. One of her favorite expressions seems to be "What if . . ." and she encourages her students to ask that question often. She assigns as much small-group work as possible and promotes class discussion as the primary method of imparting information. She emphasizes the question as much as the answer and focuses on the development of the algorithms rather than on the finished product. She considers students' answers valuable in contributing to their understanding of the situation and the process. Jane's students appear secure and confident of their ability, and they ask questions freely.

Jane is not teaching problem solving. She is teaching mathematics with a problem-solving focus. She does not need a new curriculum, but she does not hesitate to rearrange the old one. And she knows how to incorporate problem solving into the curriculum.

A good teacher of problem solving possesses the same characteristics as a good problem solver. Jane, for instance, has the *desire* to find problems, generate problems, solve problems, and evaluate solutions. She shows *enthusiasm* for the entire process and for trying new approaches in focusing on problem solving. Her enthusiasm is contagious; it captures the attention of students and colleagues alike. She has the *facility* in her subject matter and instructional procedures to make her techniques work. She is well-grounded in mathematics, heuristics, and pedagogy. She knows how to communicate problem-solving strategies and how to lead her students to develop their own methods. Finally, she has the *ability* to solve problems successfully, to generalize them, and to think of several possible methods of attack. She is comfortable with her own strengths and willing to take risks.

The Role of the Teacher Educator

Jane was not born an effective teacher nor did she become one by accident. For although the *Agenda for Action* may appear to speak primarily to classroom teachers, the responsibility for implementing its goals does not end there. Teacher educators and supervisors must also share the responsibility for teaching mathematics better. Consequently, a major challenge to teacher educators and supervisors is to discover ways to help both preservice and in-service teachers become effective teachers of problem solving. This includes showing them how to add meaningful problem-solving activities to their lessons, how to decide which traditional topics can be presented in a different order or with a different emphasis, how to communicate problem-solving attitudes to their students, how to encourage their students to see the mathematics in a wide range of situations, and how to formulate their own questions and evaluate their own answers.

Teacher education programs have traditionally recognized the importance of a knowledge of mathematics, psychology, pedagogy, and aspects of general education, and these components continue to be important. However, more is needed.

First, teachers need not only the knowledge of mathematics that is represented by their ability to perform mathematical operations and procedures but also the knowledge *about* mathematics that makes it possible for them to deliver that knowledge and understanding to pupils in a meaningful way. Second, teachers must themselves become good problem solvers, but knowledge in problem-solving techniques alone will not be enough. Since teachers are teaching a process rather than a skill, they must have solved problems themselves; they must have known firsthand the satisfaction of making some progress, the frustration of encountering temporary blocks, and the delight of solving a difficult problem. They should learn to recognize mathematical aspects of seemingly nonmathematical situations and to ask questions where previously they sought only answers. Teacher education needs to provide these experiences for the teacher.

An undergraduate methods course gave Jane her start in problem solving. Problems were assigned throughout the course, and students discussed solutions and methods of solution in class. Jane learned the importance of group work and group experience in solving problems and evaluating strategies. She realized that emphasizing the process rather than the answer made

students think about how they solved the problem so that the next problem seemed easier. Another activity in Jane's methods class that contributed to her success was collecting problems suitable for different age and ability levels. That set of problems formed the nucleus for the collection she uses now, but it contributed something even more valuable to her development as a teacher. In the collection process Jane discovered enough references and resources to produce usable problems for several years.

In addition to training teachers to solve problems and to find problems, teacher educators must be sure teachers know how to generate their own problems. One way to teach this skill is through assignments and discussions in methods classes or in-service programs. For example, the leader can assign group members the task of generating mathematical problems from nonmathematical items in newspapers or magazines. Then if each person is sent home with the same item, discussion can center on the number and types of problems developed; individuals can explain why they chose a particular aspect of the situation or why they presented it in a particular manner. A follow-up assignment could be to find other stories that lead to similar problems. Charles was in such a methods class that worked both with news items and with data collection and analysis of the results of laboratory experiments. He has no trouble generating questions from current events stories. Lately he has been thinking about a town he knows that holds a celebration every Friday the thirteenth. He has enough questions to keep his fifth graders busy for several days: How many such days are there this year? Does every year have a Friday the thirteenth? What is the most any one year can have? What is the longest time between two such celebrations? Is there a pattern in successive years? What happens during leap year?

Once teachers learn how to find and generate problems, they must also learn how to evaluate them in terms of their own students and the mathematics content of their courses. Again, a discussion approach is appropriate in developing evaluation skills. Sue and Brad, who teach twelfth and third grade, respectively, are in an in-service program for mathematics teachers in their district. The supervisor, a vibrant and imaginative problem solver himself, brings a variety of problems for the group each week. Sometimes everyone works on the same problem, and sometimes they group themselves according to teaching levels; but each session ends with a looking-back discussion. They consider how each problem could be used in a classroom, how it might be presented to a group or to an individual, how it could be adapted to different age or ability levels. They examine each problem to determine why and in what circumstances it is indeed a problem, and they decide with what mathematical topics it might be used. Finally, they discuss the heuristic techniques that could be exemplified by each problem and try to generalize the results or create an extension of each problem. Sue and Brad are each building a file of problems for their classes (sometimes they can each use the same problem in a different manner), and they have gained valuable skills in adapting and generalizing problems for their own use. They have also become adept at solving problems and, by watching the supervisor, have learned how to model problem-solving attitudes for their own students.

Just as discussion and group work are effective methods to use in teacher education programs, so are they effective with these teachers' own pupils. Consequently, mathematics teachers also need training in leading discussions and understanding group dynamics. Good mathematics teachers recognize the importance of creating a supportive and encouraging atmosphere, and this becomes even more important in teaching with a focus on problem solving. Students will not feel secure in attacking a problem unless they are confident that their attempts will be accepted and valued whether they discover a solution or merely refine a question. Thus, classroom atmosphere is another important topic for consideration in teacher education programs.

Even when preservice teachers have learned how to solve problems, find and generate problems, evaluate their solutions, and create a classroom atmosphere conducive to problem solving, they need, in addition, the opportunity to practice and observe these skills. Similarly, practicing teachers need to learn and share new ideas for implementing problem solving. Both of these needs can be accommodated by carefully placing student teachers with cooperating teachers who are chosen for their problem-solving ability, leadership, enthusiasm, and receptiveness to new ideas. As experienced teachers learn more about teaching with a problem-solving focus, they will become better able to demonstrate this facility to their student teachers. The student teachers, in turn, will bring new ideas, problems, and strategies from their mathematics education classes, and cooperating teachers who are professionally active as leaders in their schools and districts can disseminate new information to other teachers. Thus the careful placement of student teachers and the wise planning of in-service activities could eventually facilitate the incorporation of problem solving into the mathematics teaching of many others. Only then will students begin to enjoy the full opportunity to develop their problem-solving abilities.

Conclusion

The scenario of a future in which problem solving is truly the focus of mathematics instruction is exciting and challenging. However, the degree to which the *Agenda* actually becomes *action* depends on the degree to which each mathematics teacher, teacher educator, and supervisor accepts the responsibility to create that future. There is no question that we have the facility and the ability to do this and to make it stick. The real question is, do we have the desire and the enthusiasm?

2

Problem Solving: It's Never Too Early to Start

Grace M. Burton

SCHOOLS are currently being shaken by a powerful force whose rallying cry is "Back to the Basics!" Teachers of young children, as members of the mathematics education community, will do enormous good if they can assist those proponents of "basics," whose expertise is usually in other areas, to realize that there is nothing *more* basic to the intellectual development of the child than the ability to solve problems.

From their earliest years, children attempt to answer questions about their world and to order its sometimes chaotic messages. They observe, ponder, form hypotheses, collect data, test their hypotheses, and draw conclusions. In short, they set up and attempt to solve problems. This self-imposed experience is invaluable in preparing them to use similar strategies to solve problems in the world of shape, size, and quantity. It falls within the province of the teacher of early childhood classes as well as the teacher in the higher grades to provide an environment wherein these skills are fostered.

Most children enter school with a sense of wonder and a keen desire to learn. As Dewey (1933) suggested, the teacher's task is to keep alive that "sacred spark of wonder." Becoming aware that children enjoy solving problems appropriate to their developmental stage and providing enticing mental challenges will help teachers perform this special task.

Problem solving has been the subject of copious research. A review of the education literature yields much information about the characteristics of good problem solvers, some of which are an ability to note similarities and differences, to discern recurrent patterns, and to organize and interpret facts and relationships. These characteristics will be developed and strengthened by early and continued practice in problem solving.

We have also learned a great deal about the teaching of problem solving.

And what we know is as applicable to early childhood education as to the secondary level. Problems need to be interesting to the child and presented in a format he or she can decode. Thoughtful consideration rather than immediate answers should be the goal.

Many teachers associate problem solving with pages in textbooks containing "number stories." This, however, is not the only type of mathematics problem. Even children who have not yet learned to read can be presented with situations that require reflection and the generation of a unique solution rather than the recall of a learned response. At the early childhood level, appropriate problems are easily structured around three important mathematical processes—classification, seriation, and patterning.

Classification

Classification, informally known as *sorting,* involves choosing an attribute and separating a collection of objects according to whether or not they exhibit this characteristic. Not only is classification according to observable traits a prerequisite to the child's ability to form sets using more abstract criteria, it also lays the groundwork for understanding the relationships between and among sets of elements. Piaget calls classification the earliest stage of logical thinking and defines it as one of the two prerequisites for developing a true concept of number. It deserves a place in every primary classroom.

Materials for classification activities can be purchased from commercial companies, collected by teachers, or purchased at little cost from craft shops or garage sales. Commercial sets, called attribute or logic block sets, usually consist of plastic or wooden pieces in a variety of colors, shapes, and sizes. Other attributes such as thickness or texture might also be included. Less formal commercial sets are also available and consist of houses, trees, or vehicles that vary in color, size, or style.

Collected materials are among the most interesting to work with. They include seashells, door keys, dime-store jewelry, pens and pencils, toy soldiers, pebbles, leaves, tiles, play money, spools of thread, nails and screws, bottle tops, plastic picnic utensils, pictures cut from magazines, macrame beads, pattern blocks, fabric or wallpaper swatches, ceramic tiles, unshelled mixed nuts, buttons, and boxes of broken and whole crayons. A request for donations probably will yield copious quantities of material for classification. These additions to a mathematics resource shelf can be neatly stored in shoe boxes, coffee cans, plastic dishpans, or vegetable bins.

Teachers need not be highly artistic to construct a structured classification set such as that shown in figure 2.1. Wallpaper, painted wood, fabric scraps, cardboard covered with contact paper, or colored plastic lids—any of these can be used to make the shapes. Of course, a different number of shapes and

Fig. 2.1. A teacher-made attribute set

colors will work as well. For use on the overhead projector, a set made from card stock of four shapes and two sizes with zero, one, or two holes punched into each piece is both practical and effective. When the pieces are cut from transparent plastic, color becomes an additional available dimension.

Given a collection of items, most children set about sorting it, even without instruction to do so. Young children, however, often have difficulty maintaining a consistent sorting criterion. They may start to sort by color and then switch midstream to sorting by size, or vice versa. Children of five can usually sort according to color, shape, or size. Sorting by two dimensions simultaneously is very difficult for most children until age seven or eight.

Teachers who have never attempted classification as a classroom activity may believe at first that there is little more to this task than looking at the collection and moving the pieces around. Those who have been exposed to it at an appropriate level, however, will agree that it takes mental processing to discover the similarities and differences among the items in the collection and to form new collections based on a chosen attribute. Since experience is the quickest road to learning, any convinced reader might spend a few minutes trying to generate the rule used to form the subsets of the groupings in figure 2.2. It should be readily apparent that seeing alone is not sufficient.

Activities that invite children to classify can be provided for the whole class, for small groups, or for individual exploration. Teachers who wish to introduce the whole class to this process might use the song "Just Like Yours" from Hap Palmer's record *Math Readiness Songs* (used here with

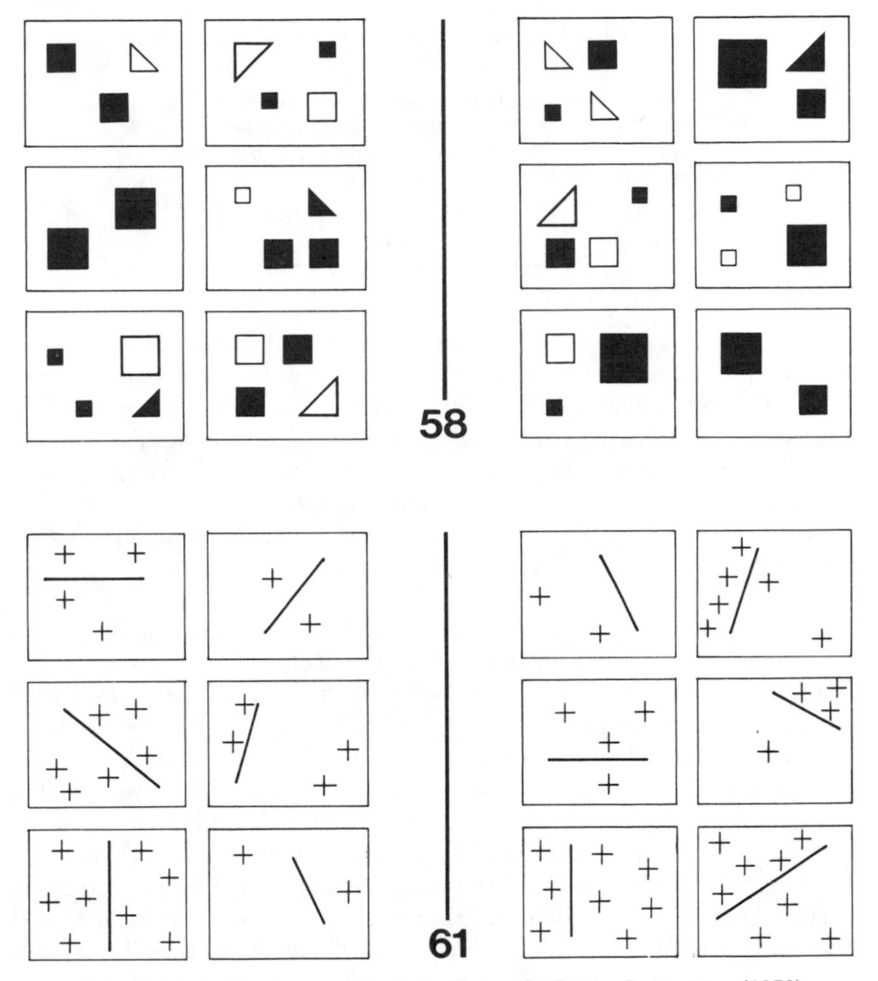

Fig. 2.2. Classification problems from Bongard's *Pattern Recognition* (1970)

permission, Educational Activities, Freeport, NY 11520). After giving each child one of three or four shapes cut from some sturdy material, the teacher might ask the children to listen carefully to the song and do what the words suggest.

Just Like Yours

Find someone who has a shape that's just like yours.
Find someone who's holding the same shape.
Find someone who has a shape that's just like yours.
Find someone who's holding the same shape.

Now trade shapes with the person you found.
So everyone is holding a different shape.

Now trade shapes with the person you found.
So everyone is holding a different shape.

Find someone who has a shape that's not like yours.
Find someone who's holding a different shape.
Find someone who has a shape that's not like yours.
Find someone who's holding a different shape.

Now trade shapes with the person you found.
So everyone is holding a different shape.
Now trade shapes with the person you found.
So everyone is holding a different shape.

Teachers can also use cut-and-paste activities to encourage classification. Often first-grade teachers teaching the "sounds" of the letters ask children to look through catalogs and old magazines and cut out the pictures of objects starting with a given letter. Although few may have thought of it that way, this activity is a classification task. Another activity would involve collecting and sorting pictures of animals that live in different habitats or examples of different modes of transportation. Although related to mathematics, these activities are important to learning in social studies or science as well.

The most engrossing of classification activities is also one of the simplest to present and can be done with an even number of children. Give each child a handful of environmental material, such as shells, and ask them to separate that material into two groups so that all the items in one group are alike in some way and all those in the other group are alike in some way and all the shells are allocated to one of the two groups (see fig. 2.3). Once everyone has finished sorting, let each child choose a partner and guess how she or he grouped the material. As this activity is repeated several times with the same material, the children will become aware that any set can be divided along several dimensions. Idiosyncratic groupings such as "the ones I like" and "the ones I don't like" or "pretty" and "ugly" may arise, as well as groups

Fig. 2.3

based on more predictable criteria such as rough and smooth, dark and light, and curved and straight. Especially effective items to use in this way are seashells, buttons, keys, and bottle tops.

Once materials have been classified, children often wish to record the groupings they have made. The most natural way is simply to put all items of one group in one pile or row and all items of the other group in another pile or row. Older children may wish to use verbal labels or more advanced symbolization to record their groupings. They might record the classification of a set of variously colored plastic spoons and forks, for example, by—

1. stacking the pieces in columns headed with cards labeled by the teacher or illustrated with pictures;
2. placing the utensils in a divided box, spoons on one side, forks on the other, and then writing on each side how many there were in that group;
3. putting items on either side of a line, tracing around them, and coloring them in;
4. gluing each object to a separate card and arranging the cards in the form of a bar graph.

To classify, children must be able to discern differences among objects and keep the identified difference in mind while considering each item of the set to be sorted. These abilities, to isolate characteristics and to evaluate instances, are important in further study. (The process can also lead to interesting discussions that build vocabulary in both mathematical and nonmathematical areas.) Doing classification activities leads a child to ponder at an appropriate level such mathematical notions as set inclusion, intersection, and the null set. It also leads smoothly to graphing, a mathematical activity that is all too often reserved for the upper grades. (Baratta-Lorton [1976] presents a wide variety of follow-up ideas on this topic.) And, of course, classification tasks present the child with a puzzling situation to which the answer is not immediately apparent—that is, a problem.

Seriation

Seriation provides a foundation for the mathematical study of inequalities. Like classification, seriation seems to be intrinsically interesting to children. As the sale of preschool toys attest, toddlers like to play with toys that nest inside each other and that can be stacked in order of decreasing size. By the time children are four, they easily seriate a set of three to five objects. They find it more difficult to seriate larger sets or to insert additional items into an already formed series.

Whereas most people think of seriation as arranging items by length or width, there are many other appropriate dimensions. Among these are height, diameter, tone, weight, and depth of hue. As well as being a second prerequisite for dealing with the more abstract ordering of numbers, seriation also gives children practice in solving nonverbal problems.

Children develop the ability to seriate efficiently the way they develop most other skills—by practice in interesting situations. The teacher can easily provide a wide variety of seriation tasks by varying the material and tailor the tasks to the needs of individual children by changing the number of items to be seriated. Some items that teachers can ask children to put in order are painted paper-towel rolls cut to different heights, identical bottles filled to different heights with plain or colored water (these can be seriated either by sight or sound or color), small boxes or baby food jars containing differing levels of sand or rice, structural materials such as Stern blocks or Cuisenaire rods, different sizes and shapes of empty cans, jar lids, paper clips, rubber balls, nails, pieces of chalk, paint chips—anything that can be distinguished by color, hue, length, width, thickness, height, or any other category at all. (see fig. 2.4.)

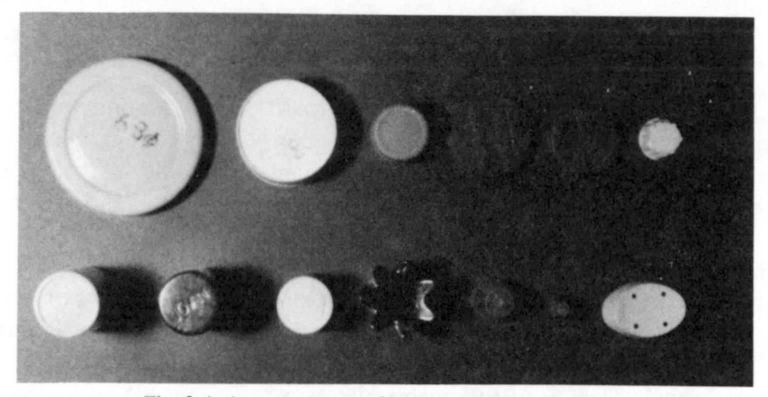

Fig. 2.4. An assortment of buttons arranged by size

Double seriation—ordering two sets of objects and then coordinating the orderings with each other—is a real challenge. (See fig. 2.5.) Another twist on seriation is to have the child seriate in the reverse order from a given set (see fig. 2.6). Reverse double seriation is the most difficult of all; it requires the child to match the largest of one set with the smallest of another. This task can challenge even a gifted child.

Experience with ordering need not be restricted to environmental material. From a catalog, cut out pictures of children of different ages and laminate them. Find a cartoon strip that tells a clear story, laminate it, and cut the frames apart. Leave these sets on a table for children to unscramble.

Providing experience in ordering will prepare the child for learning to associate numeral shapes, numeral names, and the concepts of the counting numbers. Constructing this triple abstract seriation is a much more difficult

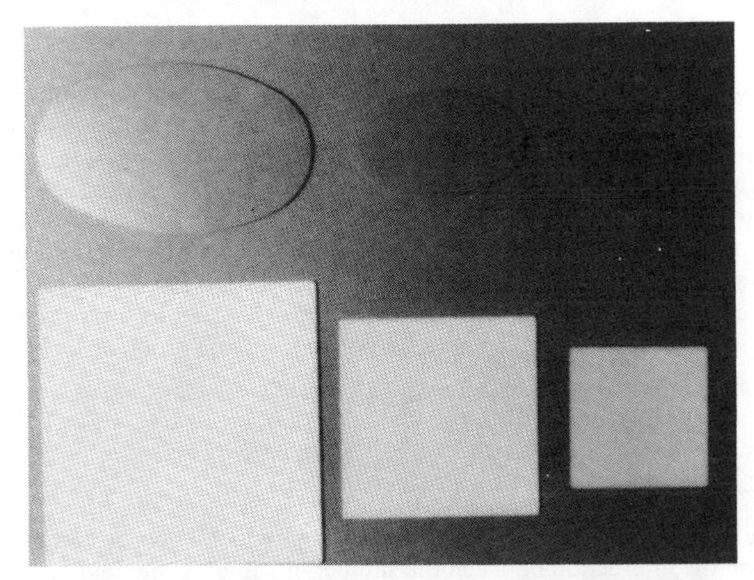

Fig. 2.5. Double seriation: ordering two sets of objects and coordinating the orderings with each other

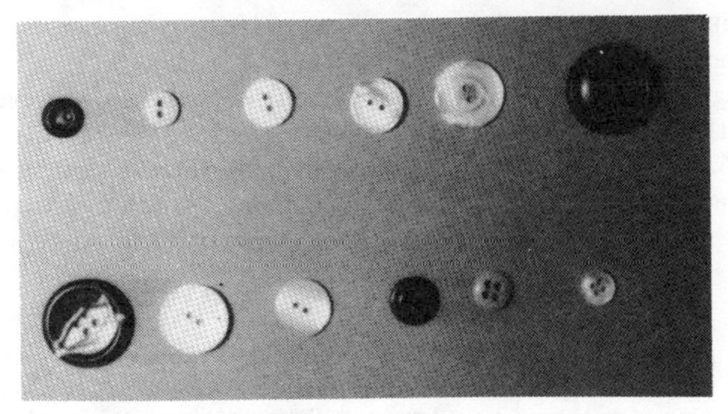

Fig. 2.6. Reverse double seriation: matching the largest of one ordered set with the smallest of another

task than many adults realize but can be eased by early practice in seriating concrete items.

It is well to remember during these activities that "the long term goals in classification and seriation are not to enable children to make little matrices and arrange little graduated sticks, but to use the *process* of classification and seriation to isolate relevant variables and to generate and test hypotheses in dealing with the real world" (Kamii 1972, p. 104). This is the essence of problem solving.

Including the problem-solving activities of classification and seriation in the early childhood classroom has an added bonus—these activities are exactly the mental processes necessary for the development of a true concept of whole number. This concept, Piaget says, "is neither a simple system of class inclusions, nor a simple seriation, but an indissociable synthesis of inclusion and seriation. The synthesis derives from the abstraction of qualities and from the fact that these two systems (classification and seriation), which are distinct when their qualities are conserved, become fused as soon as their qualities are abstracted" (Piaget 1967, p. 83). It is this fusion that is paramount in the development of the number concept. And it is attention to the two components that will help children lay a solid foundation for later mathematical endeavors, including solving those ubiquitous story problems in texts and on tests and, even more important, solving problems presented in the real world.

Patterning

Intelligence, Piaget stated, is "a structuring which imposes certain patterns on the interaction between the subject or subjects and near or distant objects" (Piaget 1950, p. 167). Recognizing a pattern requires analyzing and reflecting, noticing similarities and differences, and becoming aware of the distinctions between essential and nonessential features—thereby involving many important aspects of problem solving.

Although patterns are commonly presented using real objects, pictures, or symbols, they can also be presented with claps, drum beats, sung notes, or animal noises. It is also possible to present patterned sequences of motor activity, such as "touch your shoulders, touch your knees, touch your shoulders, touch your knees." Many circle games employ both sound and motor patterns.

Incorporating patterning into the curriculum of young children is easy and inexpensive, and it provides numerous opportunities for individual creativity. Many teachers who have first tried patterning activities hesitantly have become the most enthusiastic advocates. This is fortunate, for their classes will be exposed to a mathematical activity that stretches the mind of each student, encourages discussion as well as mental reflection, and fosters a sense of accomplishment.

When several types of patterning tasks are involved in an activity, the following should be the order of development: reproduction (copying), identification (choosing a match), extension (adding on), extrapolation (filling in a missing part), and translation (for example, changing from one color or shape to another, from a color to a shape, or from a visual presentation to an auditory response). Most four-year-olds can copy a simple pattern; extending or translating a more complicated pattern can challenge even an advanced first grader.

In a pattern, distinct elements are presented in a regular fashion, using at least three repetends. In an "AB" pattern, for example, two objects or symbols are alternated. One might have such AB patterns as gold key, silver key, gold key, silver key; dark stone, light stone, dark stone, light stone; or "woof," "meow," "woof," "meow." Some other pattern forms are ABC (do, re, mi), AABB (red, red, blue, blue), ABCD (*, !, $, ¢) and ABA (clap, snap, clap). The ABA repetend is a little more difficult for most children; there is little difference in difficulty among the others.

A patterning task involves four distinct steps: (1) the presentation of a perplexing situation to which no answer is immediately apparent, (2) the formation of a hypothesis, (3) the testing of the hypothesis, and (4) the confirmation or rejection of the hypothesis. When the solver is given time for reflection in a climate of acceptance, patterning encourages a unique and creative response (see fig. 2.6).

Once a teacher is convinced that patterning is a viable classroom activity, the next question is, "What, specifically, can I do?" Here are some suggestions:

1. Give each child a line of dots on which a pattern has been constructed (fig. 2.7). Ask them to extend or to describe the pattern (Baratta-Lorton 1976).

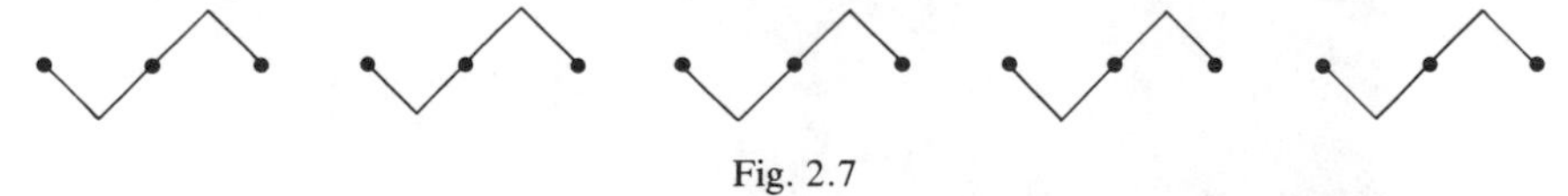

Fig. 2.7

2. Make a red, blue, green, red, blue, green train of blocks. Take out two blocks. Ask the child to fill them in.
3. Make a wall from two types of parquetry blocks, such as triangle, hexagon, triangle, triangle, hexagon, triangle. Ask the child to continue the wall or to make a wall "like it" using squares and rectangles.
4. Ask the children to join you as you alternately slap your thighs, snap your fingers twice, and clap your hands. Baratta-Lorton (1976) provides drawings of each of these motions so that children can translate the pattern they see and hear into one that can be only seen.
5. Ask children to make a pattern by tracing a simple template.
6. Perform repetitive calisthenic exercises to music. Begin the exercise and ask the children to join you as they catch on.
7. Sing musical tones according to a pattern (for example, C,C,E,C,C,E). Ask one child to echo you and then to make up a new pattern and choose an echo.
8. Give each child a necklace of colored macaroni. Ask each one to make a similar necklace using a different color. A yellow, green,

green, yellow, green, green necklace might be translated into an orange, red, red, orange, red, red necklace, for example. Children who enjoy a challenge could be asked to reverse the pattern.

9. Cut squares from a wallpaper book and ask children to find those that are identical in design but different in color.
10. Have a child make a pattern of interlocking cubes or beads and ask another to continue it, reverse it, or transpose it.
11. Act out patterns in folk dances or circle games. In "looby-lou," for example, the same motion sequence is repeated with various parts of the body. The "bunny hop" consists of a prescribed sequence of steps done over and over again.
12. Ask children to form patterns out of environmental materials; these patterns can be extended or translated into a never-ending variety of pattern sequences (see fig. 2.8).

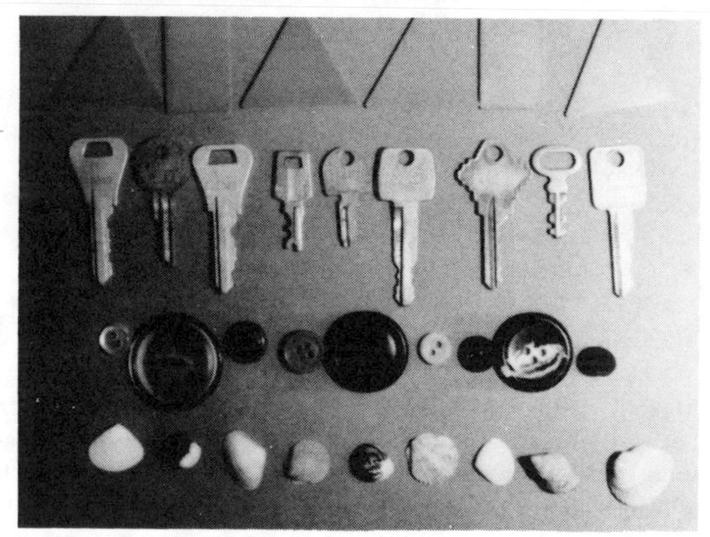

Fig. 2.8. Patterning: making a variety of pattern sequences with environmental materials

13. Make file cards with a pressure-sensitive dot sequence such as orange, purple, purple, orange, purple, purple, and then ask children to make that design with blocks of the same (or different) colors or to extend the design with other dots. Such cards make easy pattern models for those who are not artistically gifted.

Beyond the early childhood years, skill in pattern recognition will prove useful in working with sequences and series, in skip counting leading to multiplication, and in understanding the base-ten numeration system. It may not be clear at first how alternating beads on a string, making pattern block walls, or snapping interlocking cubes together prepares a child for the study of these mathematical topics or for facility in solving word problems. However, it is true that devising situations that call for a search for patterns will help students develop a level of reflective analysis beyond that attained by those who are only competent "symbol pushers."

Douglas Hofstadter, in his monumental *Gödel, Escher and Bach* (1979, p. 26), lists the essential abilities for intelligence:

To respond to situations very flexibly

To take advantage of fortuitous circumstances

To make sense out of ambiguous or contradictory messages

To recognize the relative importance of different elements of a situation

To find similarities between situations despite differences which may separate them

To draw distinctions between situations despite similarities which may link them

To synthesize new concepts by taking old concepts and putting them together in new ways

To come up with ideas which are novel

Although Hofstadter did not have it specifically in mind, these abilities also describe the abilities needed by good problem solvers—those fostered by patterning activities in the early childhood classroom.

Encouraging the Quest

In 1902, E. H. Moore asked, "Would it not be possible for the children in the grades to be trained in powers of observation and experiment and reflection and deduction so that always their mathematics should be directly connected with matters of thoroughly concrete character?" (Moore 1967, p. 367). The teacher of young children should be able to answer this question in the affirmative and to use that answer as a guide for curriculum development.

When the goal is problem solving, it is never too early to start and seldom too late to begin. What is important is that there *be* a goal and that it be recognized as worth striving for. A mathematics curriculum built on what we know about how children become good problem solvers will place minimum emphasis on pencil-and-paper busy work and maximum involvement with real materials and with situations designed to engage the child's interest. Instead of a headlong rush to process numerals or memorize facts, such a curriculum will emphasize the individual discovery and exploration of similarities and differences. Problem-solving abilities acquired at the beginning of their school careers will enable young children to be successful in their mathematical endeavors.

Piaget said, "The complete act of intelligence thus involves three essential stages: the question which directs the guess, the hypothesis which anticipates solutions and the process of testing which selects from them" (Piaget 1950, p. 94). Fortunate the student whose teachers from the beginning pose enticing problems that give rise to the first of these stages, encourage creative explorations that generate the second, and foster analytical skills that lead to the third.

BIBLIOGRAPHY

Baratta-Lorton, Mary. *Mathematics Their Way.* Palo Alto, Calif.: Addison-Wesley Publishing Co., 1976.

Bongard, M. *Pattern Recognition.* Rochelle Park, N.J.: Hayden Book Co., 1970.

Burton, Grace M. "Helping Parents Help Their Preschool Children." *Arithmetic Teacher* 25 (May 1978): 12–14.

Dewey, John. *How We Think.* Boston: D. C. Heath, 1933.

Elkind, David. *Children and Adolescents.* 2d ed. New York: Oxford University Press, 1974.

Hofstadter, Douglas. *Gödel, Escher and Bach: An Eternal Golden Braid.* New York: Basic Books, 1979.

Kamii, Constance. "An Application of Piaget's Theory to the Conceptualization of a Preschool Curriculum." In *The Preschool in Action: Exploring Early Childhood Programs,* edited by Ronald K. Parker. Boston: Allyn & Bacon, 1972.

Kofsky, Ellin. "A Scalogram Study of Classificatory Development." *Child Development* 37 (1966): 191–204.

Lee, Lee C. "Concept Utilization in Preschool Children." *Child Development* 36 (1976): 221–27.

McKillip, William D. "Patterns—a Mathematical Unit for Three- and Four-Year-Olds." *Arithmetic Teacher* 17 (January 1970): 15–18.

Moore, Eliakin Hastings. "On the Foundations of Mathematics." *Mathematics Teacher* 60 (April 1967): 360–74.

Nuffield Foundation. *Mathematics Begins.* New York: John Wiley & Sons, 1972.

Perkins, Ruth M. "Patterns and Creative Thinking." *Arithmetic Teacher* 14 (December 1967): 668–70.

Peterson, John A., and Joseph Hashisaki. "Patterns in Arithmetic." *Arithmetic Teacher* 13 (March 1966): 209–12.

Piaget, Jean. *The Psychology of Intelligence.* New York: Harcourt, Brace & Co., 1950.

———. *Six Psychological Studies.* New York: Random House, 1967.

3

Focusing on Problem Solving in the Primary Grades

Francis (Skip) Fennell

ONE of the challenges of focusing on problem solving is to carefully involve younger pupils. The strategies presented here consider the learning needs of pupils in the primary grades (K–3), recognizing that one of the major difficulties in teaching problem solving to younger children is identifying activities that consider both pupil readiness and curricular goals. Problem-solving activities should be included within all areas and levels of the primary grade mathematics curriculum. That is, pupils should be confronted by activities that cause them to critically examine and use the skills and concepts they learn in daily mathematics lessons. Problem solving can be an integral portion of the typical mathematics lesson. There is no questioning the importance of developing and reinforcing such typical primary grade skills as counting, place value, addition, and subtraction. These skills are not only useful but critical for more advanced mathematics topics. However, there is often a tendency to overemphasize the mastery of computational skills. This reduces the possibility of including problem solving as a viable topic in the primary grades.

Just how does problem solving fit, at this level, into a learning philosophy that monitors pupil readiness, emphasizes a manipulative-based concept-development approach to learning, and also includes a concern for computational skills? Problem-solving activities should cause primary grade pupils to explore, translate, verbalize, and verify. Problem-solving experiences will extend and expand their knowledge to meet the demands of the problem-solving setting. The most important issue to consider here is the consistent integration of problem solving into the primary grade mathematics program.

Strategies for Problem Solving in the Primary Grades

Let's turn our attention to specific strategies for teaching problem solving at the primary level. The strategies suggested here are easily used on a daily basis. They are general in focus to allow their use with a myriad of K–3 mathematics topics, and they include a variety of teaching options that offer a high level of pupil involvement in creating and solving problems. These strategies can ensure a consistent approach to the development of problem-solving ability at the primary level.

Oral examples

An early introduction to problem solving using oral examples is encouraged. Ask children a variety of questions relating to both introductory and reinforcement mathematics activities:

1. How many are wearing red today?
2. How many will be buying milk today?
3. How many are in the Jets reading group?
4. How many children are in the class?
5. How many more boys are there than girls in the class?
6. How many pairs of sneakers are being worn today?

Be sure to integrate oral problems with each day's mathematics activities.

Preparing for word problems

Two related strategies that are useful for introducing pupils to word problems are the mathematics-experience approach and pupil group discussions. Both strategies focus on pupil interaction and the careful, gradual development of thinking strategies based on teacher and pupil questioning. The class or small-group focus of these strategies provides an excellent opportunity for pupils to validate a solution by modeling it with manipulatives.

The mathematics-experience approach is a problem-solving strategy that has its roots in the language-experience approach of the reading class. Each day the teacher elicits a problem from the class, copies it on large chart paper (fig. 3.1), and then has the class, through questioning and discussion, solve the problem. With certain classes and levels the teacher may want to formulate the problem and involve the class only in its solution. The class should be involved in discussing procedures for solving these problems. Using the problem-solving staircase technique illustrated in figure 3.2, the teacher has pupils respond to each step in the staircase as they complete the problem.

Possibilities for using the mathematics-experience approach include word and picture problems relating to numeration, addition, subtraction, measurement, probability, fractions, multiplication, and division. This class-oriented approach is an effective way to begin the day by focusing on problem solving.

The pupil group-discussion strategy focuses on discussion as a way to articulate solutions to word problems. Students at the primary level often

Brett had 2 dogs and 3 cats. He found 6 more cats. How many cats does he have now?

Fig. 3.1

have difficulty or lack confidence in their ability to solve problems. This group-discussion strategy presents problems at a slower pace with a major goal being to involve the group in their solution.

1. **What's the Problem?**
   What is known?
   What is unknown?
2. **Make a Plan.**
   How will you solve the problem?
3. **Do It!**
   Carry out your plan.
4. **Check Your Work.**
   Is the answer reasonable? (Does it make sense?)

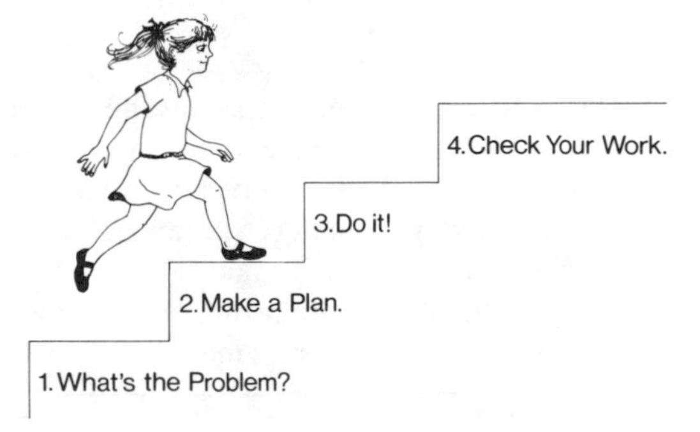

Fig. 3.2

Questioning and logic activities

Strategies that involve pupils through questioning may include logical-thinking activities. These activities sequence the information revealed to pupils in a problem. Teachers should provide information on paper strips to be taped on the chalkboard or reveal it line by line using an overhead projector. Some examples of this type of logical-sequence problem appear in figure 3.3.

Have the group suggest possible solutions to the exercises, but limit them to ten responses. Vary the activity by having pupils find a random number between 0 and, say, 100 by asking questions that the teacher can answer with

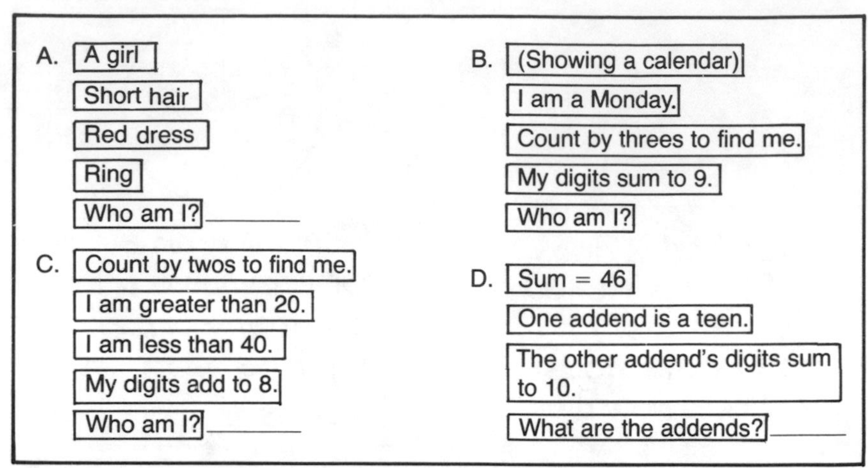

Fig. 3.3

yes or no. See how many questions are needed before the number is identified. Logical-sequence activities provide additional problem-solving examples that truly involve pupils in questioning and responding. They are fun!

Pictures to words

The pictures-to-words strategy helps move primary grade pupils from iconic, or pictured, representations to actual word problems. This strategy is an excellent diagnostic procedure for involving even remedial students in problem-solving activities. Students should be presented with pictured sequences, such as those in figure 3.4, and respond first verbally and then in writing to help form the problem itself and then find its solution. Primary grade pupils will enjoy making and solving their own picture strips.

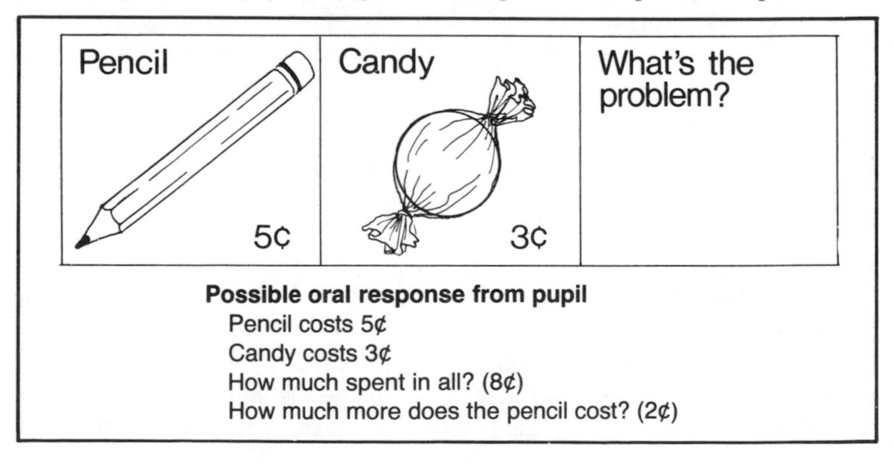

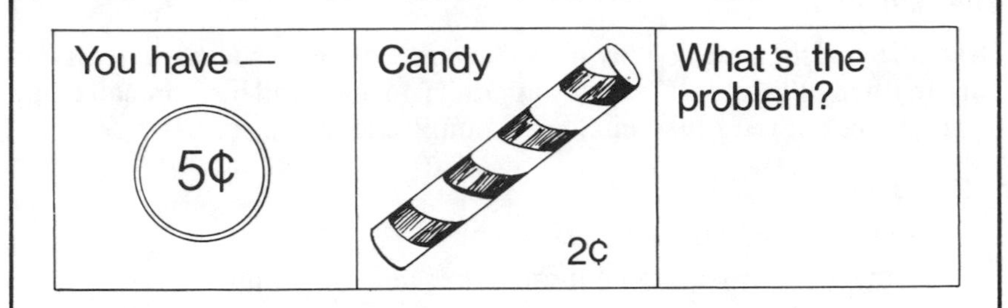

Possible oral response from pupil
You have 5¢
Candy costs 2¢
How much change?

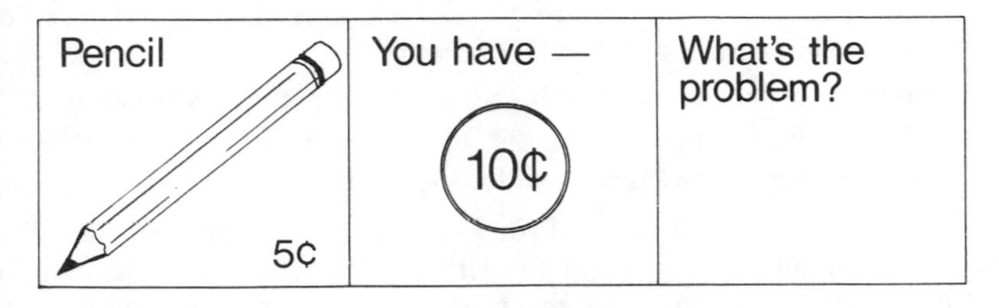

Pupils write out problem statements:

____

____

____

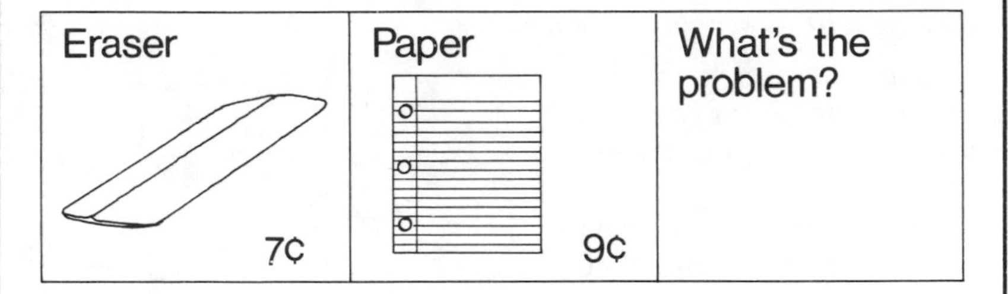

Pupils write out problem statements:

____

____

____

Fig. 3.4

The math drama

Dramatize mathematics? Why not! The strategy here is to have pupils act out a mathematical story. Take any appropriate topic and have a small group of pupils act out the situation. Some examples follow.

Counting

1. Brett bounced his tennis ball 8 times.
2. Jenny gave crayons to all those in her reading group.
3. Ask the ninth student in line to run around the room.

Operations

1. Seven students were sitting. How many were boys and how many were girls? (+)
2. The Greatest American Hero visited the school. He knocked over 3 of the 8 chairs in front of the classroom. (−)
3. Mark gave each member of his reading group 2 pieces of candy. (×)
4. Stacey collected 12 hats from the class. She put them in 2 closets. Each closet had the same number of hats. (÷)

This activity can really be fun if you tell your "actors and actresses" that they must not talk but only act out the activity. Then have the class guess the problem that was presented. A larger class project may be to select a favorite short nursery tale and revise it to include mathematics problems. "The Three Little Pigs" is easily adapted by including questions about the cost or quantity of building materials used in home construction as well as about huffing and puffing! The math drama is a perfect after-lunch activity.

Math newsletter—school and home

Communicating to others about problem solving can be an enjoyable learning experience. Having each class or instructional group submit a problem to the school's problem-solving newsletter further strengthens a school's commitment to problem solving. Consider sharing this newsletter with parents. Students will enjoy attempting to solve problems created by other classes and grades. Parents may be encouraged to assist their children in completing the problems in the newsletter.

Applications with numbers

Primary grade pupils need to see how numbers can be used in applied situations. The newspaper, store catalogs, and, to some extent, the telephone book provide excellent examples of applied problem-solving situations. Primary mathematics topics involving time and money provide for further applications. Sample applications appear below.

Telephone book. Use the telephone book and find a number with a sum over 50; under 20. Use the class list of telephone numbers and determine whose number has the greatest and least sum.

Catalog. Have pupils estimate the cost of a variety of commonly advertised catalog items. Initially consider toys, popular appliances, clothing, and furniture. Sequence the articles to be estimated according to pupil awareness. Have pupils check the reasonableness of their estimates by cutting out and mounting actual illustrations and costs from the catalog. Determine the best catalog estimators. Repeat this activity to develop estimation skills.

Newspaper. Have pupils clip and paste ten different monetary amounts from the newspaper. The amounts should be ordered from the least to the greatest. A variation of this comparison activity would be to review the paper's weather report and circle the ten highest and ten lowest temperatures for a particular day.

Additional applications involving catalogs, telephone books, and newspapers might include the following:

- Locate different sections of the newspaper using the paper's index.
- Recognize ordinal and cardinal numbers in headlines and stories.
- Find and cut out different geometric shapes.
- Determine the vowel count in a particular newspaper story. Graph the results.
- Determine the cost for your family to go out to a movie, sports event, or dinner.
- Determine the difference between regular and sale prices.
- Make a shopping list of items whose total cost is less than, say, $100.
- Interpret and create different types of graphs.
- locate fractions and decimals in the newspaper. Cut out and order them.
- Cut out as many measurement terms, abbreviations, and symbols as you can find.
- Use the evening TV schedule to compare timed amounts.

Thinking math

The "thinking math" strategy is an attempt to challenge primary grade pupils with nonroutine types of problem-solving activities. These examples should stimulate many students, but not all students will successfully complete them. Analyzing a challenging problem of the week may help develop "great thinkers" in primary grade classrooms. Try these examples:

A. 3, 5, 8, __?__
What number goes in the blank? Why?

B. How many ways can you make change for 50¢?

Calculator

The calculator should be consistently used in the primary grades. It can be useful in solving and checking a variety of activities. As a class treat, have pupils use calculators for solving all problems on Friday. Make sure they recognize that the calculator will help only in the *do it* and *check your work* steps of the problem-solving staircase. They will still need to determine *how* to solve the problems.

Try these strategies in your primary grade classroom. They will fit most instructional topics, and your children will be the real winners as you focus on problem solving.

BIBLIOGRAPHY

Fennell, Francis M. *Elementary Mathematics: Priorities for the 1980's.* Bloomington, Ind.: Phi Delta Kappa Educational Foundation, 1981.

Krulik, Stephen, and Jesse A. Rudnick. *Problem Solving—a Handbook for Teachers.* Boston: Allyn & Bacon, 1980.

Polya, George. *How to Solve It.* Princeton, N.J.: Princeton University Press, 1971.

4

An Instructional Approach to Problem Solving

Richard Brannan
Oscar Schaaf

PROBLEM SOLVING has been a subject of research by mathematics educators, educational psychologists, and philosophers since the 1930s. Yet tests, including the recent mathematics tests of the National Assessment of Educational Progress, indicate that our school students and even graduates have serious deficiencies in problem-solving ability. An examiniation of current mathematics texts and the results of classroom visitations reveal that problem solving is only a minor part of the mathematics instruction in both elementary and secondary schools.

To alleviate this deficiency, the Oregon Department of Education Title IV-C office funded the Problem Solving in Mathematics (PSM) project in 1977. Housed at the Lane Education Service District, the project staff has developed mathematics materials to be used in grades 4–9. What follows are a definition of a problem, some basic assumptions, and the instructional approach used by the PSM staff to implement Recommendation 1 of NCTM's *Agenda for Action,* namely, that *problem solving be the focus of school mathematics in the 1980s.*

What Is a Problem?

Suppose Sean, a sixth grader, is asked to look at the drawing in figure 4.1 and fill in the missing output blanks for *a, b,* and *c* in the table. Would this be a problem for him? Probably not, since he would only need to follow the directions.

Suppose Shelley, a second-year algebra student, were asked to consider

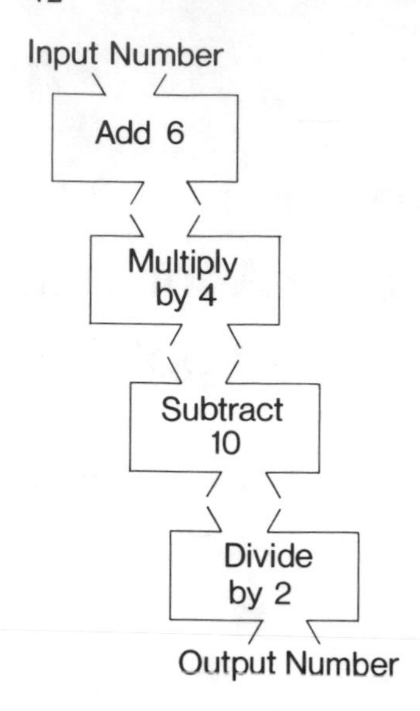

	Input Number	Output Number
a.	4	
b.	8	
c.	12	
d.		39
e.		47
f.		61

Fig. 4.1

the same drawing and fill in the missing input blank for *d.* Would this be a problem? Probably not, since she would write the suggested equation

$$\frac{4(x + 6) - 10}{2} = 39$$

and then solve it for the input.

Now suppose Sean, the sixth grader, were asked to fill in the input for *d.* Would this be a problem for him? It probably *would* be. He has no directions for getting the answer. However, if he has the desire, it is within his power to find the answer. What might he do? Here are some possibilities:

1. Sean might *make a guess, check the guess,* and then *make refinements* until he gets the answer.

2. He might fill in the output numbers that correspond to the input numbers for *a, b,* and *c* (fig. 4.2) and then observe this pattern:

If the input is increased by 4, the output is increased by 8.

Such an observation should lead quickly to the correct input of 16.

	Input	Output
a.	4	15
b.	8	23
c.	12	31
d.		39

Fig. 4.2

3. He might start with the output and work *backward* through the machine hookup using the inverse (or opposite) operations.

For Sean, there was no ready-made way to find the answer, but most motivated sixth-grade pupils would find a way.

A problem, then, is a situation in which an individual or group accepts the challenge of performing a task for which there is no *immediately obvious way to determine a solution.* Frequently, the problem can be approached in many ways. Occasionally, the resulting investigations are nonproductive. Sometimes they are so productive that they lead to many different solutions or suggest more problems than they solve.

Some Basic Assumptions

Pupils can improve their problem-solving abilities if they are able to—

- identify and discuss the problem-solving skills they use while solving mathematical problems;
- solve problems independently;
- make sense out of the mathematics they are learning.

New content being learned must be an extention of previous mathematical understandings.

Teachers can help pupils improve their problem-solving abilities if they—

- teach a variety of problem-solving skills directly;
- use a problem-solving approach to mathematics instruction frequently with a variety of mathematical activities;
- are themselves problem solvers who share their experiences (successful and unsuccessful) with pupils;
- create a classroom atmosphere in which openness and creativity can occur.

An Instructional Approach to Problem Solving

Yes, pupils become better problem solvers if they learn a variety of problem-solving skills and have opportunities to apply the skills. To fulfill the assumptions above, this instructional approach is advocated:

1. *Teach problem-solving skills directly.* Early in the year use direct instruction to teach such tactics as *guess and check, look for a pattern,* and *make a systematic list.* Spend ten to fifteen minutes a day on problems that are written specifically to highlight particular problem-solving skills.

2. Use problem-solving activities throughout the school year. Provide lessons in each of the following areas:
   a) Drill and practice
   b) Laboratory-type of investigations
   c) Mathematical concept development

3. Have pupils use their problem-solving skills on nonroutine or challenge problems. The creative meshing of several skills to form a solution strategy is the most crucial phase of problem solving.

4. Develop an open and creative atmosphere:
   a) Set an example by actually solving problems yourself.
   b) Encourage communication and cooperation with the problem-solving process.
   c) Use your pupils' ideas in problem solving.

Integration of Materials into the Curriculum

The integration of appropriate materials for pupils into the regular curriculum is most important. Pupils must come to understand that problem solving is an everyday part of the mathematics class (as well as life!) and not an add-on topic for two weeks. The following are examples developed by the PSM staff. In the materials, each page has an accompanying teacher commentary printed on the back.

1. The following are examples to teach problem-solving skills.
   a. Guess and check.

Each row, column, and diagonal of this magic square adds to 15.

Use the numbers 2, 3, 5, 6 8, and 9 to complete the magic square.

		4
7		
	1	

   b. Look for a pattern.

1 x 8 + 1 = ______

12 x 8 + 2 = ______

123 x 8 + 3 = ______

1,234 x 8 + 4 = ______

a. Predict the answer for 123,456 x 8 + 6. ______

b. Check your prediction.

c. Predict and check 12,345,678 x 8 + 8. ______

c. Make a systematic list.

The coins shown above are the only ones you have.
What amounts can you make if you use 1, 2, 3, or 4 of the coins?

d. Make a drawing.

A fireman stood on the middle step of a ladder.
As the smoke got less, he climbed up three steps.
The fire got worse so he had to climb down five steps.
Then he climbed up the last six steps and was at the top of the ladder.

How many steps were in the ladder?

e. Make a reasonable estimate.

Megan is at a material shop. She has a $10 bill. Which of these purchases can she make?

a. 3 yards of material at $2.98 a yard.

b. A scissors for $8.15 plus a spool of thread at 85¢.

c. A pattern for $2.75 plus 4 yards of material at $2.10 a yard.

d. Four sets of buttons at $1.25 a set and 2 balls of yarn at $1.95 a ball.

f. Eliminate possibilities.

Ms. Ashley has less than 100 pieces of candy.

If she makes groups of 2 pieces, she will have 1 piece left over.
If she makes groups of 3 pieces, she will have 1 piece left over.
If she makes groups of 4 pieces, she will have 1 piece left over.
If she makes groups of 5 pieces, she will have no pieces left over.

How many pieces of candy could she have?

2. Figures 4.3, 4.4, and 4.5 show examples of drill-and-practice activities. They are used throughout the school year but are particularly useful at the time a concept is being reviewed or practiced.

3. Examples of the laboratory type of investigations can be seen in figures 4.6, 4.7, and 4.8. Taken primarily from geometry and probability, the activities usually require active participation by students and extra hands-on material.

4. Each grade level contains sections to introduce a new concept through a problem-solving approach. Figure 4.9 shows the first and fourth and figure 4.10 the fifth and eighth lessons of a sequential unit to develop multiplication and division concepts for fourth graders. The unit uses a rectangular grid as a model for these concepts. Other topics presented in this manner are fractions, decimals, percent, equations, and graphing.

5. Later in the school year, students are given opportunities to use their skills to solve challenge problems. The instruction is essentially nondirective. The teacher's role is to help students understand the problem and to listen, praise, encourage, give hints, and summarize or emphasize solution methods. Some sample challenge problems appear in figures 4.11–4.13.

The chart in figure 4.14 shows all the sections of each of the grade-level packets in the PSM materials. The curriculum can be organized around problem solving.

Research on the PSM Project

The summaries that follow are paraphrased from reports by Dizney, Lovell, and Mittman (1980), Peacock (1979), and Mayes (1980). The report of Dizney et al. was based on data from a test developed by the Problem-Solving Skills Test (PSST) project and from the 1978 Metropolitan Mathematics Survey Test.

> On the PSST, substantial gains occurred for both fourth- and fifth-grade students. The differences were statistically significant for the total group. In addition, all seven fourth-grade classes and six of nine fifth-grade classes showed significant improvement.
>
> On the Metropolitan Mathematics Survey Test, students overall progressed at a somewhat faster rate than expected. Four of seven fourth-grade classes and seven of nine fifth-grade classes improved their standings relative to the nationally normed instrument.

[*Text continues on page 59*]

DIGIT SHUFFLE

1. This problem uses the digits 1, 4, 7, and 9. Complete the problem.

   $\begin{array}{r} \boxed{9}\,\boxed{4} \\ \times\ \boxed{1}\,\boxed{7} \\ \hline \end{array}$

2. Use the same digits. Fill in the blanks. Try to get an answer larger than 6000.

   $\begin{array}{r} \square\,\square \\ \times\ \square\,\square \\ \hline \end{array}$

   Find one other.

   $\begin{array}{r} \square\,\square \\ \times\ \square\,\square \\ \hline \end{array}$

3. Find two problems with answers smaller than 1000.

   $\begin{array}{r} \square\,\square \\ \times\ \square\,\square \\ \hline \end{array}$ $\begin{array}{r} \square\,\square \\ \times\ \square\,\square \\ \hline \end{array}$

4. Find an answer close to 4000.

   $\begin{array}{r} \square\,\square \\ \times\ \square\,\square \\ \hline \end{array}$

5. Each multiplication problem uses the digits 6, 5, 3, and 2. Circle the problem you predict would give the larger answer.

   63 x 52 or 62 x 53

   Check your prediction. Discuss your conclusions.

Fourth Grade

A NUMBER TIMES ITSELF

Use a calculator. Fill in the table.

14 x 14	
24 x 24	
34 x 34	
44 x 44	
54 x 54	
64 x 64	
74 x 74	
84 x 84	
94 x 94	

Now solve these problems. <u>In each case the number in the two boxes must be the same.</u>

1. $\square \times \square = 1225$
2. $\square \times \square = 361$
3. $\square \times \square = 3249$
4. $\square \times \square = 4225$
5. $\square \times \square = 784$
6. $\square \times \square = 9216$
7. $\square \times \square = 6724$
8. $\square \times \square = 11025$

<u>EXTENSION</u>

$\square \times \square \times \square = 17576$

The number in each box must be the same.

Fifth Grade

Fig. 4.3

TEST ANSWERS

Harriet's test answers are shown below. Estimate only to decide which answers are not correct. Mark an X beside the wrong answers.

1. 39.82 + 5.76 = 45.58
2. 2.38 + 6.5 + 9.27 = 12.30
3. 3.024 + 5.718 + 11.135 = 19.877
4. 6.23 + 16.84 + 19.21 = 42.28
5. 17.91 + 30.2 + 23.62 = 44.55
6. 14.98 - 2.76 = 12.22
7. 8.392 - 5.14 = 3.252
8. 32.23 - 29.96 = 17.73
9. 62.824 - 39.98 = 37.164
10. 39.84 - 26.83 = 13.01
11. .3 x 932 = 279.6
12. 6 x 25.6 = 153.6
13. 3.9 x 14 = 54.6
14. 2.3 x 5.2 = 119.6
15. 4.32 x 8.37 = 3615.84
16. 8.324 ÷ 4 = 2.081
17. 92.8 ÷ 8 = 11.6
18. 35.25 ÷ 15 = 2.35
19. 18.24 ÷ 3.2 = .57
20. 480.33 ÷ 59.3 = .81

---

1. Work the ones Harriet missed to find the correct answers.

2. Explain why she missed the ones she did.

Sixth Grade

CREATE A PROBLEM

Write a single digit in each square to create a correct problem. Digits may be repeated in a problem.

1. □□□ + □□□ = 561
2. □□□ + 873 = □□□□
3. □9 + 7□ + □□ = 2□□
4. □□□ - 589 = □□□
5. 5284 - 9□□ = □□□□
6. □□□□ - □4□ = 293
7. □□ x 8 = □□□
8. 195 x □□ = □□□□□
9. □□□ x □ = 3372
10. □□□ x 29 = □□□□1
11. □ ⟌ □□□ = 27
12. □ ⟌ □□□□ = 1234 R 5
13. 53 ⟌ □□□□ = □□□
14. □ ⟌ 1234 = □□□
15. □□ ⟌ □□□□ = 23 R 31

Create your own problems. Give them to a friend to solve.

Seventh Grade

Fig. 4.4

A TRICKY PATTERN

Study this pattern: 3, 4, 7, 11, 18, 29, 47, 76

Note that 3 + 4 = 7
4 + 7 = 11
7 + 11 = 18
etc.

Use the same rule to complete the patterns below.

1. 4, 5, ___, ___, ___, 37

2. 1, 2, ___, ___, ___, ___, 21

3. 2, 1, ___, ___, ___, ___, 18

4. 6, 10, ___, ___, ___, ___, 110

5. 2, ___, 10, ___, ___, ___, 74

6. 3, ___, ___, 7, ___, ___, ___, 50

7. 5, ___, ___, 5, ___, ___, ___, ___, 65

8. 6, ___, ___, 4, ___, ___, ___, 35

9. 8, ___, ___, ___, 7, ___, ___, ___, 41

10. ___, ___, ___, ___, 15, 20

11. ___, ___, ___, ___, ___, ___, ___, ___, 10, 16

12. 10, ___, ___, ___, ___, 0, ___, ___, ___, ___, 10

Eighth Grade

DRAWING GRAPHS

Draw these graphs.

1. a. The graph is a straight line.
   b. It goes through the origin.
   c. It goes through the point $(^{-}3,5)$.

2. a. The graph is a circle.
   b. The origin is the center of the circle.
   c. It passes through the point $(^{-}10,0)$.

3. a. The graph is a straight line.
   b. It is parallel to the x-axis.
   c. The y-coordinate is always $^{-}3$.

4. a. The graph is a curved line.
   b. The graph is only in the first quadrant.
   c. It passes through: (2,12) (8,3) (6,4) (12,2).

5. a. The graph is a straight line.
   b. It slopes sharply up from lower left to upper right.
   c. It passes through $(^{-}2,^{-}6)$ and (2,6).

6. a. The graph is a straight line.
   b. The x-coordinate is always equal to the y-coordinate.
   c. It passes through the origin.

7. a. The graph is a straight line.
   b. It slopes up from left to right at a 45° angle.
   c. It passes through a point with coordinates $(0,^{-}1)$.

8. a. The graph is a straight line.
   b. It is parallel to the y-axis.
   c. The point (4,0) is one unit to the left of the graph.

Ninth Grade

Fig. 4.5

TANGRAM SHAPES - 2

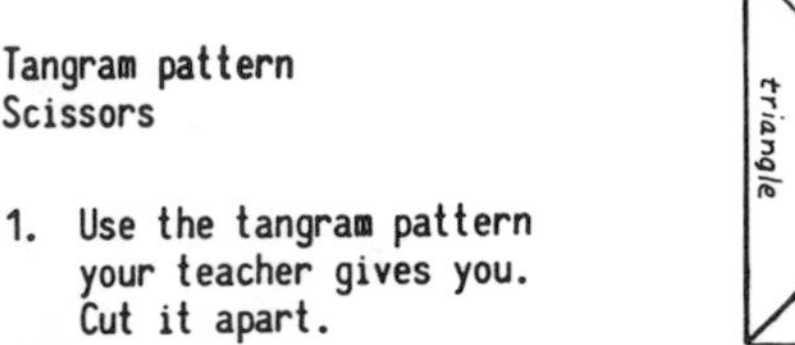

GET Tangram pattern
Scissors

DO 1. Use the tangram pattern your teacher gives you. Cut it apart.

Use the two small triangles and the parallelogram. Make:

. a long rectangle . a larger triangle
. a larger parallelogram

Sketch and label each shape.

2. Use the two small triangles, the square and the parallelogram. Make:

. a long rectangle . a large parallelogram
. a 6-sided shape

Sketch and label each shape.

3. Make many different shapes using all five triangles. Sketch each shape.

WHICH DO YOU THINK WILL BE LARGER?

1. Two players need three regular dice. Player A gets two dice. Player B gets the other one.

2. Do this activity.

   a. Player A rolls both dice and finds the product of the two numbers.

   b. Player B rolls the one die and multiplies the number by itself.

   c. The winner is the player with the largest answer.

   Who do you think will win more often? ______

3. Do the activity 30 times. Record the results in a table showing three things: A's answer, B's answer, and the winner.

Number	1	2	3	4	5	6	7	8	9	10
A's Answer										
B's Answer										
Winner										

Number	11	12	13	14	15	16	17	18	19	20
A's Answer										
B's Answer										
Winner										

Number	21	22	23	24	25	26	27	28	29	30
A's Answer										
B's Answer										
Winner										

EXTENSION

Do the activity this way:

1. Player A adds the two numbers.
2. Player B adds the number to itself.

Now who do you think will win more often? ______

PAPER, SCISSORS, OR ROCKS

Needed: 3 players

Do you know the paper, scissors, or rocks game? All players make a fist and together count to four. On the count of four, each player shows either

a. paper by showing four fingers

b. scissors by showing two fingers

c. rock by keeping the fist together.

1. Play this game with two partners. Decide who is

   Player A ________ Player B ________ Player C ________

2. Play the game 25 times with these rules:
   a. Player A gets a point if all players show the same sign.
   b. Player B gets a point if only two players show the same sign.
   c. Player C gets a point if all players show a different sign.

3. Tally the winning points in this table.

Player	Tally	Total
A		
B		
C		

4. Is this a fair game? ____ Which player would you rather be? ____

5. Make a list of the ways three players could show the signs.

6. What change in the points awarded will make the game more fair?
7. Play the game 25 times with your rules. Record the results.

Sixth Grade

BOXING UP CUBES

Kenny and Mike decided to start a part-time business--manufacturing one-inch cubes for classroom use. Now they need to decide how to box them up.

Each box is to contain 100 cubes. The problem is--which size box will use the smallest amount of cardboard?

2"
5"
10"

Kenny drew a picture like the one above. Will it hold 100 cubes?

1. How much cardboard is needed? (In other words, what is the total area?) Don't forget to include the top.

2. Determine the dimensions of all boxes that have volumes of 100. Also, find the total surface area of each. (Do not use decimals.) Place your results in the table below.

Volume	Dimensions	Total Surface Area
100	5 by 10 by 2	
100		
100		
100		
100		
100		
100		
100		
100		
100		

3. Which size box will use the smallest amount of cardboard?

EXTENSION Suppose you wanted to design a box that would hold 100 cubic centimetres of cereal. Determine the dimensions that use the least amount of cardboard. Use decimals. A calculator would be helpful.

Seventh Grade

Fig. 4.7

SNIPPING QUADRILATERALS

Needed: Centimetre ruler
Scissors

Follow these directions:

- . Draw a large quadrilateral with all sides of different lengths.
- . Find the middle point of each side.
- . Connect the midpoints as shown in the figure.

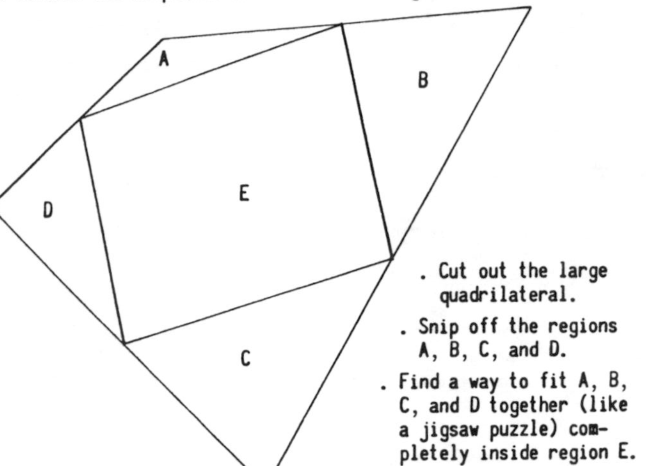

- . Cut out the large quadrilateral.
- . Snip off the regions A, B, C, and D.
- . Find a way to fit A, B, C, and D together (like a jigsaw puzzle) completely inside region E.

1. Write exact directions as to how you were able to fit the pieces together.
2. Experiment with another case. Do your directions still work? If not, change them so they do.

<u>EXTENSION</u> Sammy started with a square. He discovered that it was possible to fold regions A, B, C, and D over so as to completely fill region E. Will this folding method work for any other quadrilaterals? Explain.

RECTANGLES IN ALGEBRA

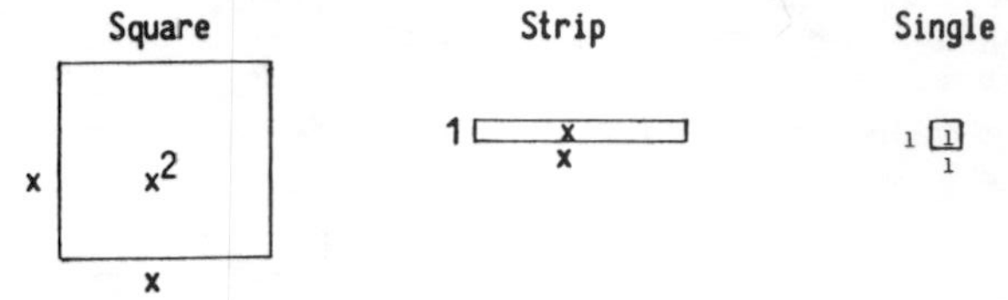

1. Get 1 square, 5 strips, and 6 singles. This can be written as $x^2 + 5x + 6$. Use the pieces to make a rectangle.

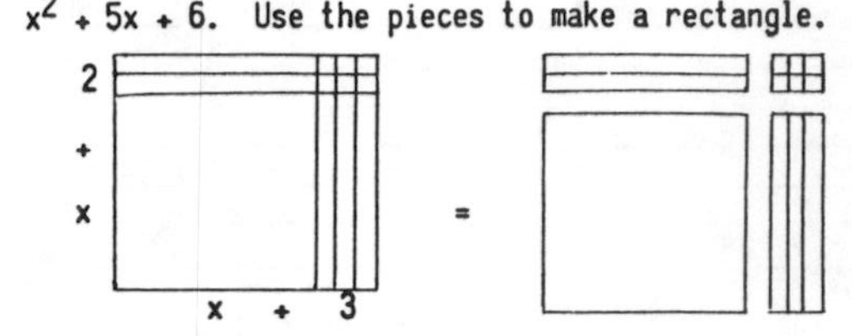

This shows the algebraic statement:

$$(x + 2)(x + 3) = x^2 + 3x + 2x + 6 = x^2 + 5x + 6.$$

2. Use the number of squares, strips, and singles shown. Make a rectangle. Write an algebraic statement describing the rectangle.

   a. 1 square, 7 strips, 6 singles

   b. 1, 7, 12

   c. 1, 6, 9

   d. 1, 9, 18

   e. 2, 5, 2

   f. 4, 8, 3

MAKING RECTANGLES

1. Take some tiles.
   Make this 4 by 3 rectangle.

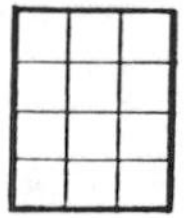

   It is 4 tiles high.
   It is 3 tiles wide
   It uses ____ tiles in all.

2. Take 16 tiles. Make as many different rectangles as you can that each use 16 tiles.

   Use the squared paper below. Record each rectangle.
   Shade in any rectangles that are also squares.

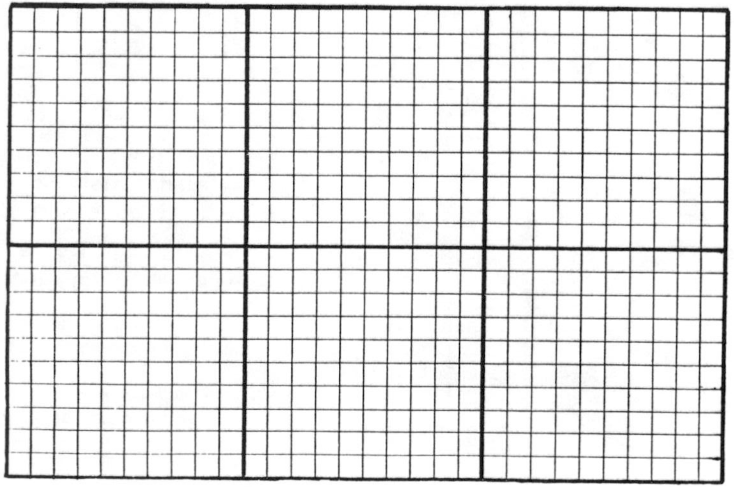

BUILDING RECTANGLES

1. Use 20 tiles. Build a rectangle that is 5 tiles high.

   How wide is the rectangle? ________

2. Use tile. Build these rectangles and give the missing dimensions.

a.	24 tiles	4 high	____ wide
b.	36 tiles	6 high	____ wide
c.	32 tiles	8 high	____ wide
d.	25 tiles	5 high	____ wide
e.	28 tiles	7 high	____ wide
f.	40 tiles	5 high	____ wide
g.	42 tiles	6 high	____ wide
h.	18 tiles	____ high	2 wide
i.	20 tiles	____ high	5 wide

Fig. 4.9

DIVISION MODELS

1. This rectangle shows that $5 \times 7 = 35$.

   It also shows that $35 \div 5 = 7$.

   Use the drawing to explain this.

2. Write a division fact for each rectangle below. Fill in any missing dimensions.
   The first problem is done for you.

a.

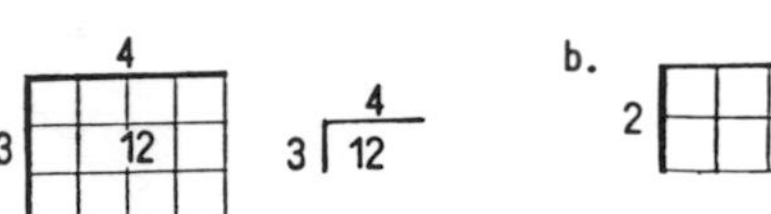

b.

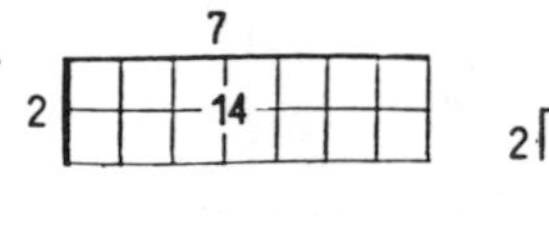

c.

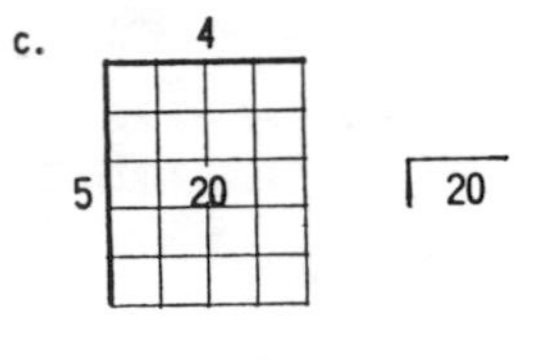

d.

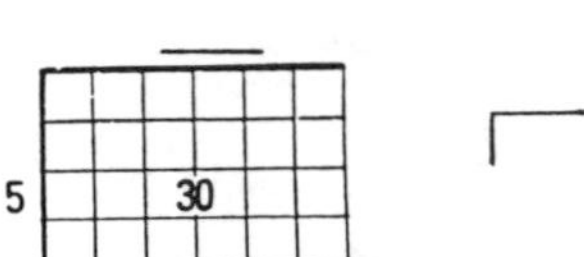

e.

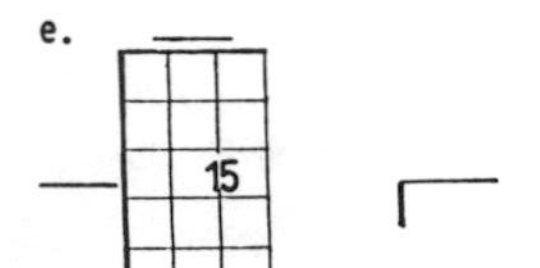

f.

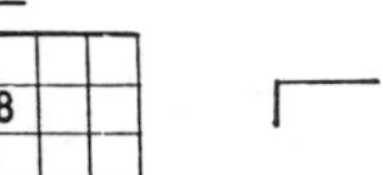

WRITTEN RECORDS

Use tile. Solve $4\overline{)29}$

Record on squared paper.

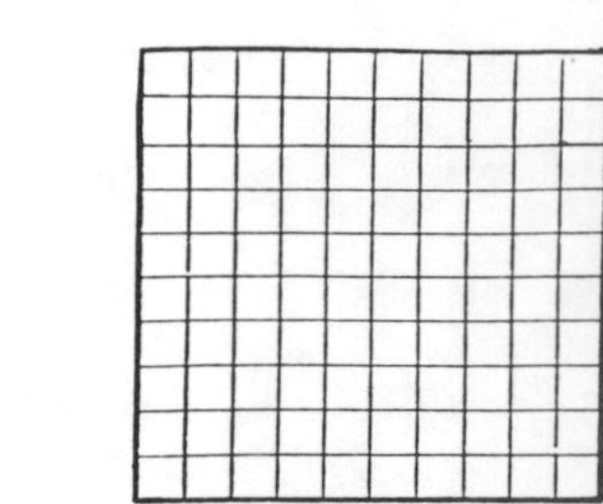

This problem can be written

$$\begin{array}{r} 7 \\ 4\overline{)29} \\ -28 \\ \hline 1 \end{array} \quad \text{Remainder } 1$$

Can you explain the written record? Discuss.

1. Complete the written record for each drawing below.

a.

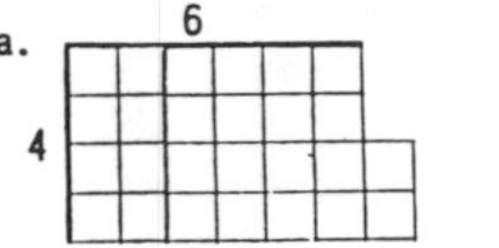

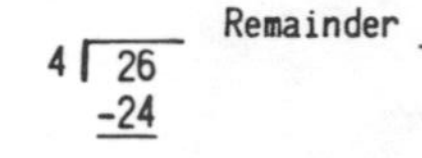

$4\overline{)26}$
-24

Remainder ____

b.

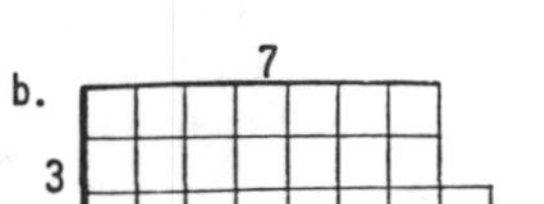

$3\overline{)22}$ Remainder ____

c.

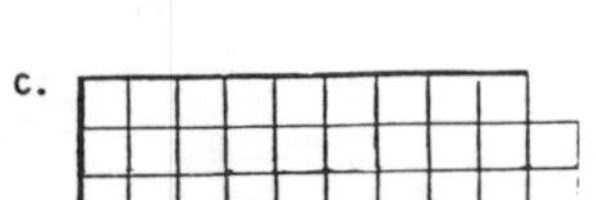

Remainder ____

PIRATE PATTY'S TREASURE

1. Pirate Patty protects her treasure chests by placing large cement blocks around the chest. (On _1_ treasure chest ___ blocks.)

Block	Block	Block
Block	Treasure Chest	Block
Block	Block	Block

2. She always places new treasure chests end-to-end.

Treasure Chests

_2_ treasure chests

___ blocks

3. Treasure Chests

_3_ treasure chests

___ blocks

4. _4_ treasure chests

___ blocks

5. _5_ treasure chests

___ blocks

6. How many blocks for:
   a. 10 treasure chests? ___
   b. 15 treasure chests? ___
   c. 100 treasure chests? ___

Fourth Grade

DUCKS AND COWS

1. Farmer McDonald raises ducks and cows.

   The animals have a total of 9 heads and 26 feet.

   How many ducks and how many cows does Mr. McDonald have?

Try to solve these "ducks and cows" puzzles. If a puzzle is not possible, explain why.

2. 9 heads and 20 feet
3. 10 heads and 24 feet
4. 8 heads and 18 feet
5. 9 heads and 50 feet
6. 6 heads and 17 feet
7. 10 heads and 18 feet.

EXTENSION

Farmer McDonald raises ducks and cows. He is out standing in his field and sees some of each kind of animal. Altogether he sees 24 feet (not including his own!). How many ducks and how many cows does he see? Show all possible answers.

Fifth Grade

Fig. 4.11

ADDITION CHALLENGE

1. Use the digits 1 through 9 once each. Fill in the circles to make the sum of 999.

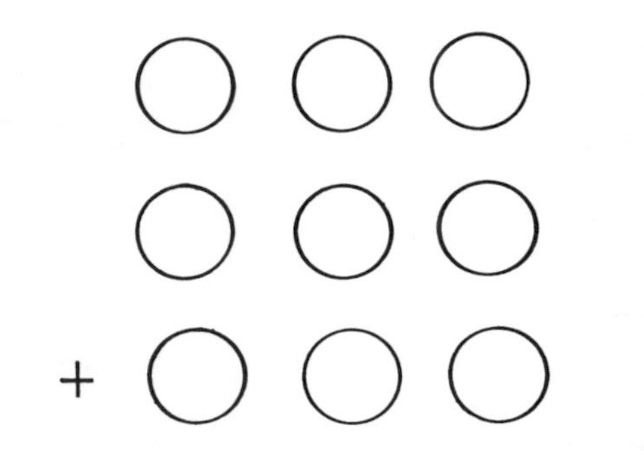

2. Find other solutions. Record.
3. Is it possible to find a solution without using 1 in the hundreds column?
4. Use the digits 1 through 9 once each in the circles to make the sum 777.

Sixth Grade

MONEY IN THE BANK

Jane and Arthur each have $10 in the bank.

Every month Jane plans to add $1 to her account.

Arthur plans to add $3 to his account every month.

1. In how many months will Arthur have twice as much as Jane?
2. Now solve the problem if they each start with $15 rather than $10.
3. Solve the problem if they each start with $20.
4. Write about anything you've discovered.

<u>EXTENSION</u>

If they each start with $10, when will Arthur have three times as much as Jane?

Seventh Grade

Fig. 4.12

THE LAST DIGIT

5^{1000} is a huge number.

In fact, the answer has close to 700 digits.

Do you think it is possible to determine the last digit of 5^{1000} ?

TRY IT.

Find the last digit of these number giants:

1. 6^{1000}
2. 11^{1000}
3. 9^{1000}
4. 7^{1000}
5. 17^{1000}

<u>BONUS</u>

What are the last two digits of 6^{1000} ?

Eighth Grade

WINDOW PANES

Sally works in a window factory. Her job is to program the computer. It tells the warehouse how many window panes to send to the assembly line.

Window panes come in three types:

a. corner panes
b. edge panes
c. center panes

A 3 by 3 window is shown.
It uses 4 corner panes,
4 edge panes, and 1 center pane.

1. Make a drawing and a table to show the number of panes of each type needed for these windows:

   a. 2 by 2 b. 3 by 3 c. 4 by 4 d. 5 by 5 e. 6 by 6

2. Study the table. How many panes of each type are needed for an <u>n</u> by <u>n</u> window?

3. Would the factory ever need window panes like these: ☐ or ☐ ?

4. The factory received a large order for rectangular windows, 2 by 3, 2 by 4, 2 by 5, etc. Help Sally decide how many panes of each type are needed. Include a 2 by <u>n</u> window.

5. Investigate 3 by <u>n</u> windows, 4 by <u>n</u> windows, 5 by <u>n</u> windows, and <u>m</u> by <u>n</u> windows.

Ninth Grade

Fig. 4.13

PROBLEM SOLVING IN MATHEMATICS

Grade 4	Grade 5	Grade 6	Grade 7	Grade 8	Grade 9
Getting Started	Getting Started	Getting Started	Getting Started	Getting Started	Getting Started
Place Value Drill and Practice	Whole Number Drill and Practice	Drill and Practice	Drill and Practice-Whole Numbers	Drill and Practice	Algebraic Simplification
Whole Number Drill and Practice	Story Problems	Story Problems	Drill and Practice-Fractions	Variation	Algebraic Explanations
Multiplication and Division Concepts	Fractions	Fractions	Drill and Practice-Decimals	Integer Sense	Equation Solving
Fraction Concepts	Geometry	Geometry	Percent Sense	Equation Solving	Word Problems
Two-digit Multiplication	Decimals	Decimals	Factors, Multiples, and Primes	Protractor Experiments	Binomials
Geometry	Probability	Probability	Measurement-Volume, Area, Perimeter	Investigations in Geometry	Fractions
Rectangles and Division	Estimation with Calculators	Challenges	Probability	Percent Estimation	Graphs and Equations
Challenges	Challenges		Challenges	Calculator	Graph Investigations
				Probability	Systems of Equations
				Challenges	Challenges

Fig. 4.14

Peacock's study showed that—

- students were using the problem-solving skills they had been taught by their mathematics teachers with purpose and direction;
- students gave indications of emotional involvement (elation, disappointment, enthusiasm, relief, interest, and determination) as well as intellectual involvement while solving problems;
- on several occasions, students referred to problem solving as being an opportunity to "use my brain."

Mayes's study indicated that—

- overall, students learned problem-solving skills to a satisfactory extent;
- the development of basic numeracy skills was not negatively affected by using the materials;
- teachers said their students enjoyed working with the materials;
- no negative or disparaging remarks about the materials were heard from students.

Summary

Among the steps recommended in *An Agenda for Action* for making problem solving the focus of mathematics education are these three:

1. The mathematics curriculum should be organized around problem solving.
2. Appropriate curricular materials to teach problem solving should be developed for all grade levels.
3. Mathematics teachers should create classroom environments in which problem solving can flourish.

Motivational materials, appropriate for a given grade level, are the most important component. The samples shown in this article are just a few of the approximately ninety activities for each grade level. The goal of the PSM project is to build competence in using a problem-solving approach to teaching mathematics. With the materials and appropriate in-service training, the goal can be accomplished and Recommendation 1 can be realized.

REFERENCES

Dizney, Henry, Byrne Lovell, and Arthur Mittman. *Evaluation Report for an Innovative Grant Title IV-C Project: Prepare and Evaluate Mathematics Problem Solving Kits for Grades 4–9.* Unpublished report submitted to the Oregon Department of Education, September 1980.

Mayes, Leslie. "An Evaluative Study of the First Year of an Elementary School Mathematics Problem-solving Program." Unpublished doctoral dissertation, University of Oregon, December 1980.

Peacock, Alistair A. "Analyses of Process Sequence Traces Observed in Mathematical Problem Solving." Unpublished doctoral dissertation, University of Oregon, September 1979.

5

Eureka! A Problem-solving Course for the Pittsburgh School District

Howard G. Bower

THE summer of 1979 presented a challenge to the staff and teachers of the Pittsburgh schools. Under the threat of a court order to integrate the school system, and hoping to achieve integration voluntarily, the staff planned a new magnet school program for September. (A magnet school is one with an "exclusive" program that is designed to draw students voluntarily from other areas of the city for the purpose of achieving voluntary integration.) To meet the needs of newly integrated students coming from other schools, a special enrichment course was designed to accompany a mathematics laboratory course for both a math-science magnet and an engineering-architecture magnet.

The mathematics specialist decided that a course devoted to problem-solving techniques was needed. Two such courses were designed: a one-semester course meeting five days a week and a full-year course meeting twice weekly. These courses were intended for students who had completed one year of both algebra and geometry and who indicated an interest in a mathematics or science career. A committee of teachers was charged to develop a course stressing problem solving with an emphasis on engineering problems.

"Eureka" was introduced to the math-science magnet school during the 1980–81 school year. The course was taught two days each week; a computer-science course was offered for the remaining three school days. The teachers devised a program organized into six units.

Each unit consists of an introduction, objectives, suggested activities, projects, enrichment activities, evaluation, and references. No single textbook was adopted for this course; instead, a variety of texts was selected for use by the teachers. (A list of course materials follows this article.) The classroom is equipped with nine TRS-80 microcomputers, a terminal connected by phone lines to a central computer, books, materials, and supplies for making projects.

Unit 1 strives to develop the students' interest in problem solving and to indicate the varied types of problems they might encounter throughout the year. It gives an overview of the course and includes thinking exercises to be done alone, problem solving for a group, and experiments for students in a mathematics laboratory.

The specific objectives for Unit 1 are—

- to develop self-confidence in problem solving;
- to work as a group to extrapolate the unknowns, the data needed, and the course of action for a problem;
- to design a project with certain specifications.

Thirty-four problems are given to the students for discussion. Some examples are these:

- Why can't a man's nose be 12 inches long? (*Answer:* Because then it would be a foot.)
- Can a person have 5 heads? (*Answer:* Yes, if you can count the *fore*head.)
- If you have only one match and entered a room in which there was a kerosene lamp, an oil stove, and a wood-burning stove, which would you light first? (*Answer:* The match.)

The class then works together in solving problems. For example, how many different objects can be constructed using one eraser, one pencil, and one paper clip? If you had $500 to construct a deck next to the family room, how would you plan and complete the task? After working together on different problems, the students are assigned exercises or enrichment activities to work on in the mathematics laboratory.

Units 2 and 3 present techniques that have proved successful in attacking certain types of problems and give the student basic tools for solving similar or related problems.

The objectives for Unit 2 follow:

- To read and restate the problems
- To list the given information and partial solutions that occur with the initial reading
- To identify the assumptions that are being made
- To construct an organized list of numbers that meet the conditions of the problem
- To tabulate specific examples that meet the conditions of the problem
- To draw a diagram that illustrates the conditions of the problem
- To construct a physical model that represents the conditions of the problem
- To identify the different cases involved in a problem

The teacher presents several problems and involves the students in brainstorming techniques. Some of the problems are "fun" and some are more mathematical. For example, students might work on the problem shown in figure 5.1 in the mathematics laboratory:

Each of four pieces of cardboard is either blue or orange on one side and has either a square or a trapezoid on the other side. Laid on a table, they look like this:

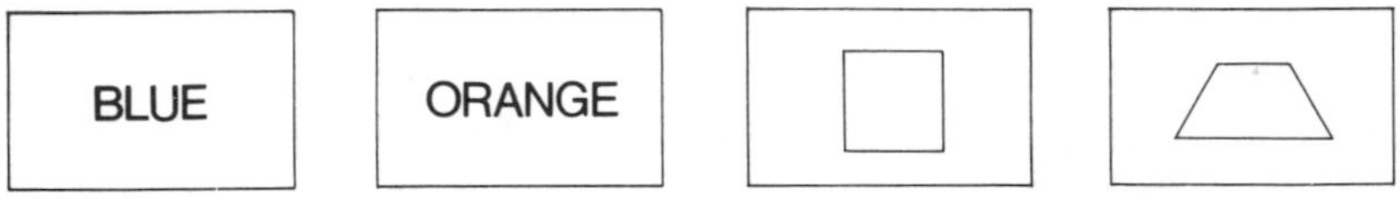

BLUE | ORANGE | □ | ⏢

Which one(s) must you pick up and turn over in order to have sufficient information to answer the question:

Does every blue one have a ⏢ on the other side?

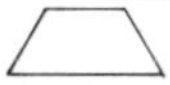

Fig. 5.1

The teacher might use the worksheet in figure 5.2 for reviewing geometry and developing the students' ability to work with sequences. The students could be given a week-long problem such as the "Tower of Hanoi" as an evaluating test for Unit 2.

UNIT 2 WORKSHEET
Pattern Solving

Name: ______________ Date: ________ Score: ________

Complete each of the following exercises and try to write a formula for the nth term in the sequence that evolves.

1. DIAGONALS OF A POLYGON (n = number of vertices)

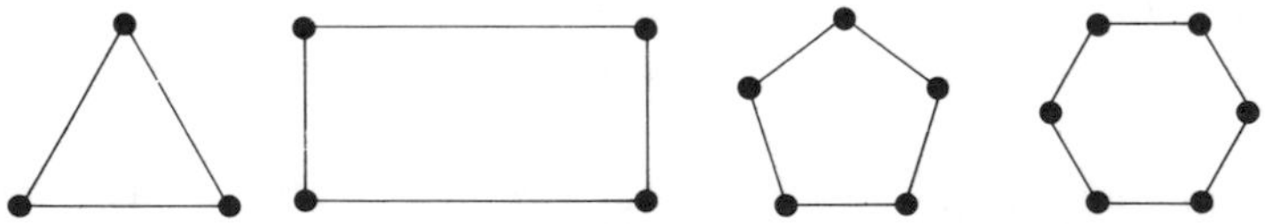

MAXIMUM NUMBER OF DIAGONALS (D) THAT CAN BE DRAWN

n	3	4	5	6	7	... 50 ...	n
D							

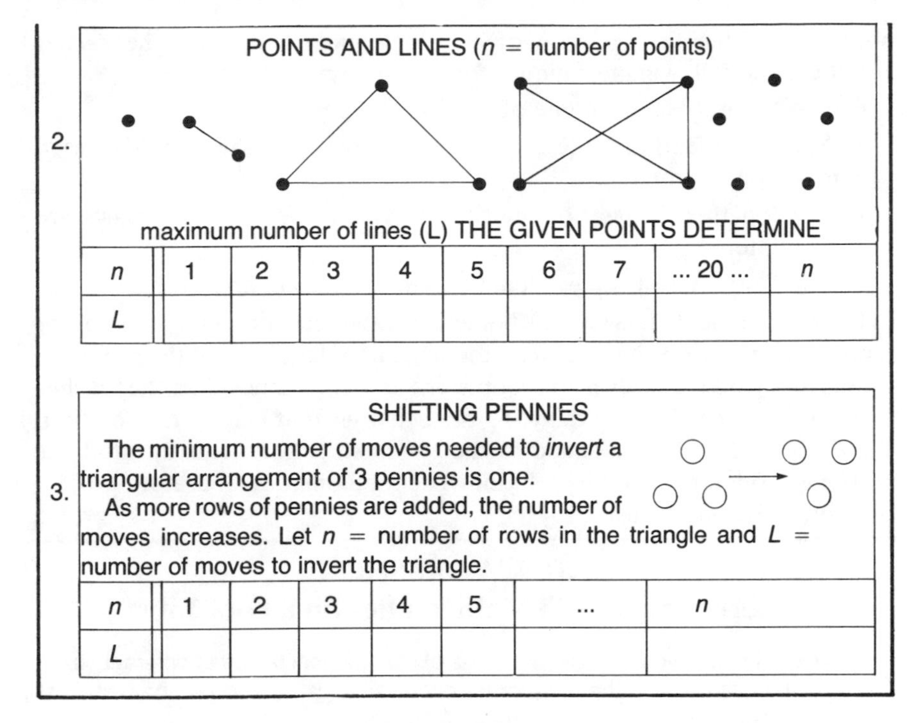

2\. POINTS AND LINES (n = number of points)

maximum number of lines (L) THE GIVEN POINTS DETERMINE

n	1	2	3	4	5	6	7	... 20 ...	n
L									

3\. SHIFTING PENNIES

The minimum number of moves needed to *invert* a triangular arrangement of 3 pennies is one.

As more rows of pennies are added, the number of moves increases. Let n = number of rows in the triangle and L = number of moves to invert the triangle.

n	1	2	3	4	5	...	n
L							

Fig. 5.2

The objectives for Unit 3 are the following:

- To develop patterns when solving problems with numerical sequences
- To draw diagrams when counting is involved in problem solving
- To distinguish the component parts of a problem using the deductive method
- To develop an organized list using the deductive process
- To design a table using deductive reasoning
- To illustrate given facts using Venn diagrams
- To design a model of the conditions in the problem

The students continue to work problems individually as part of a group. One project involves constructing with string all the possible rectangles having perimeters equal to the given length of the string. Other activities include using tangrams.

In Unit 4 the student practices problem-solving strategies. The problems are grouped in five categories by mathematical content rather than problem-solving activities. The teacher discusses each problem in relation to the techniques covered in previous units.

The objective for Unit 4 is for students to increase their problem-solving abilities in arithmetic, algebra, geometry, logic, and enrichment. The teacher could use the following techniques for group discussion:

- Select a problem from one of the five categories.
- Ask the students to suggest techniques or approaches that could be used to solve the problem.
- Develop their suggestions until the problem is solved or flaws are encountered.
- If student interest wanes, demonstrate the correct approach.

For independent classwork or homework, the teacher could select problems dealing with one or several mathematical topics. Some of the problems in this unit lend themselves to group work as well. Rather than the teacher leading the discussion and directing the activities that lead to the solution, the problem could be assigned to the class as a whole with the students working cooperatively on its solution. The mathematics laboratory project in figure 5.3 could be used.

UNIT 4 WORKSHEET
Experimenting with Mathematics on a Pool Table

This experiment involves scale drawings of special pool tables of different sizes. We are interested only in the four corner pockets: *A, B, C,* and *D.* The ball must always be shot along a 45° path from the *A* pocket.

The teacher discusses and demonstrates mathematics on a 3×4 pool table.

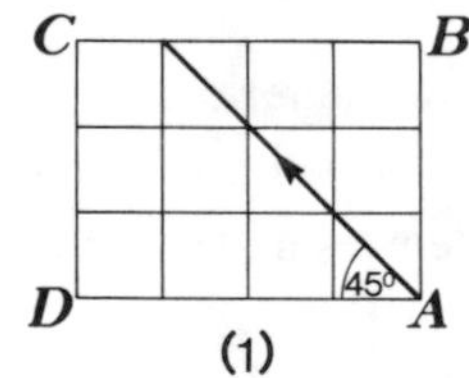

(1)

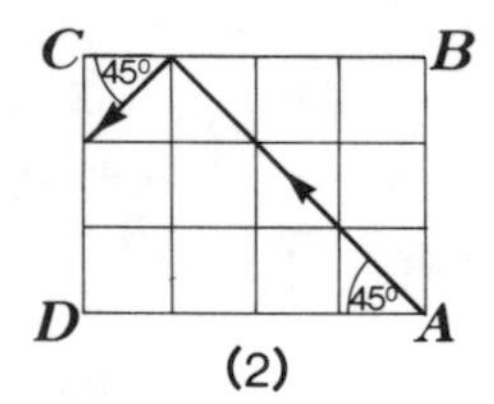

(2)

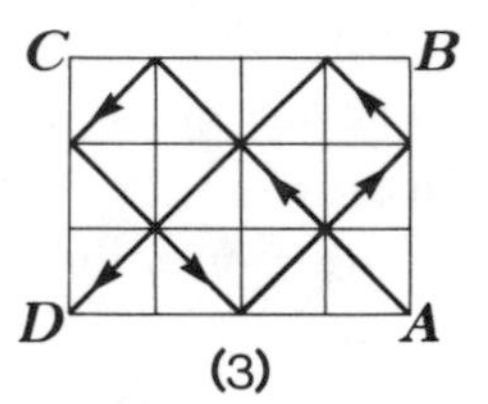

(3)

Questions: Into which pocket will the ball drop? (*D*)
How many bounces will the ball make before dropping into a pocket? (*5*)

The students should be provided with a data table and record their results after the 3×4 pool table is discussed:

Size of pool table		Number of Bounces	Ball Drops into Pocket	Answers
Width	Length			
3	4	5	*D*	
5	6			9 D
3	8			9 D
3	5			6 C

3	7			9 C
5	7			10 C
4	5			7 B
2	3			3 B
2	5			5 B

Having completed the data table for the various pool table sizes, the teacher tells the students to examine it carefully. There is a definite relationship between the number of bounces (b) and the dimensions of the table (l, w). Students are asked to discover the "Bounce Rule" ($b = l + w - 2$) when at least one dimension is *odd.*

Students should also discover the "Pocket Rule":

If w is *odd* and l is *even, D* pocket
If w is *odd* and l is *odd, C* pocket
If w is *even,* and l is *odd, B* pocket

Two additional problems for students are these:

- In trying to help a student who did not have enough points to pass the semester, a teacher offered to give the pupil a chance to earn fifty extra-credit points. On ten index cards the teacher wrote the number 50, and on the other ten she wrote nothing. Giving the child all the cards and two boxes, the teacher told him to divide the cards into two sets and to place one set in each box. With the student's eyes closed, the teacher mixed the boxes and the contents of each box. At random the child picked one box and one card out of that box. If he picked a card marked 50, the student would have enough points to pass the course. How should the cards be divided between the two boxes to maximize the probability of drawing a card marked 50? What is that probability? (Answer: One 50-point card in box A and all other cards in box B. The probability is $14/19 \doteq 3/4$.)
- Multiple choice questions can be very difficult. Can you deal with the following?

  Grsmr kxcgob sc mybbomd?

  a) Ylfsyecvi xyd drsc yxo.

  b) Pybqod sd.

  c) lye kbo k qyyn qeoccob.

  d) Kxydrob sxknoaekdo kddowzd.

  Hint: Grsmr kxcgob sc mybbomd?
  Which answer is correct? (Answer: *c*)

Fig. 5.3

The class should be encouraged to participate in the interscholastic mathematics leagues and contests held in the area during the school year. Unit 5 should be introduced as early as possible if the class is planning to compete because it is designed to help students expand their ability to solve typical contest problems. The objectives for Unit 5 are to help students—

- expand their ability to solve mathematical contest problems;
- discover a reasonable approach to the problem, given the correct answer;
- increase their problem-solving techniques by analyzing a detailed solution to a problem they were unable to solve and transferring this knowledge to related problems;
- develop their ability to take standardized tests;
- identify their areas of weakness in mathematics and concentrate on overcoming them.

The students could devise a mathematics contest for their school or city and set up the tournament so that it would produce only one winning team. In addition to local and regional contests, groups of students could also participate in the American (formerly "Annual") High School Mathematics Examination.

Unit 6 is designed to help students apply their knowledge of problem-solving techniques to the fields of engineering and architecture. The activities lend themselves to class discussion and research, followed by student experimentation or model building. Creativity is of the utmost importance in this unit.

The objectives for Unit 6 are these:

- To analyze a problem and identify the areas of research needed
- To identify materials needed for the design or construction of a given problem, and experiment with these materials
- To develop critical thinking skills leading to creative modifications in the design of a project
- To perceive the importance of mathematics in the physical world

Some of the projects that students may construct are mousetrap-powered cars, egg drops, balsa wood bridges, wind-powered cars, and paper airplanes.

During the year the students visit scientific and business organizations, participate in a mathematics league, and send teams to mathematics contests. All students take the American High School Mathematics Examination.

TIME SCHEDULE

		Report period in which unit is to be covered	Weeks in which unit should be covered	Number of days allotted to cover unit material	Number of days allotted for evaluation	Total number of days
Unit 1	Setting the Stage	1	1–3	5	1	6
Unit 2	Basic Skills for Problem Solving	1	4–8	10		10
Unit 3	Successful Problem-solving Techniques	2	9–13	10		10
Unit 4	Practice Problem-solving Techniques	2, 3	14–21	14	2	16
Unit 5	Preparation for Contest Participation	3, 4	Option 1 22–29	16		16
		2, 3, 4	Option 2—it is suggested that the teacher begin to work with material from this unit as soon as Unit 3 is completed. This material will complement Unit 4.			
Unit 6	Engineering and Architectural Problems and Projects	4	Option 1 30–35	12		12
		3,4	Option 2—it is suggested that the teacher begin work with material from this unit as soon as Unit 4 is completed. These projects are complicated and will require considerable research and experimentation outside the classroom.			

Course Materials

Basic texts

Greenes, Carole, John Gregory, and Dale Seymour. *Successful Problem Solving Techniques.* Palo Alto, Calif.: Creative Publications, 1977. (one per teacher)

Salkind, Charles T. *Annual High School Mathematics Examinations,* vol. 1. Washington, D.C.: Mathematical Association of America, 1961. (one per pupil)

———. *The Contest Problem Book II.* Washington, D.C.: Mathematical Association of America, 1966. (one per pupil)

Salkind, Charles T., and James M. Earl. *The Contest Problem Book III.* Washington, D.C.: Mathematical Association of America, 1973. (one per pupil)

Supplementary texts

Andree, Josephine, and Richard V. Andree. *Cryptarithms.* Norman, Okla.: Mu Alpha Theta, 1978. (one per teacher)

Ball, W. W. Rouse, and H. S. M. Coxeter. *Mathematical Recreations & Essays.* Toronto: University of Toronto Press, 1974. (one per teacher)

Barnard, Douglas St. Paul. *Adventures in Mathematics.* New York: Hawthorn Books, 1965. (one per teacher)

Butts, Thomas. *Problem Solving in Mathematics.* Glenview, Ill.: Scott, Foresman and Company, 1973. (one per teacher)

Carroll, Lewis. *Pillow Problems and a Tangled Tale.* New York: Dover Publications, 1958. (one per teacher)

Charosh, Mannis. *Mathematical Challenges II plus Six.* Reston, Va.: The National Council of Teachers of Mathematics, 1974. (one per teacher)

Fixx, James. *Games for the Superintelligent.* Garden City, N.Y.: Doubleday & Co., 1972. (one per teacher)

Gardner, Martin. *New Mathematical Diversions from "Scientific American."* New York: Simon & Schuster, 1971. (one per teacher)

Graham, L. A. *Ingenious Mathematical Problems and Methods.* New York: Dover Publications, 1959. (one per teacher)

Hepler, Donald E., and Paul I. Wallach. *Architecture, Drafting and Design.* New York: McGraw-Hill Book Co., 1971. (one per teacher)

Hess, Adrien I. *Mathematics Projects Handbook.* Reston, Va.: National Council of Teachers of Mathematics, 1977. (one per teacher)

Krulik, Stephen. *A Mathematics Laboratory Handbook for Secondary Schools.* Philadelphia: W. B. Saunders Co., 1972. (one per teacher)

Polya, G. *How to Solve It.* Princeton, N.J.: Princeton University Press, 1973. (one per teacher)

Posamentier, Alfred S., and Charles T. Salkind. *Challenging Problems in Algebra I.* New York: Macmillan Co., 1970. (one per teacher)

Spitler, Gail. *Mathematics in Modules, MP1.* Skokie, Ill.: Rand McNally & Co., 1976. (one per teacher)

———. *Mathematics in Modules, MP2.* Skokie, Ill.: Rand McNally & Co., 1976. (one per teacher)

Summers, George. *Mind Teasers: Logic Puzzles & Games of Deduction.* New York: Sterling Publishing Co., 1977. (one per teacher)

Vergara, William C. *Mathematics in Everyday Things.* New York: American Library of World Literature, 1962. (one per teacher)

Equipment to be ordered

Calculators (10 per class)

Microcomputers (2 per class)

Quiz-a-Matic 10 player model #249
University Research Co.
7581 Palos Verdes Dr.
Goleta, CA 93017 (805/968-9852)

Dominoes (double 6, double 9, double 12) (5 of each per class)

Balsa wood

Construction paper

Scissors

String

Straws

Glue

Toothpicks

Graph paper

Compasses

Rulers

Note: Because proposing problems is an ongoing activity, the teacher should be able to purchase materials needed to construct and design the models used in analyzing given problems.

BIBLIOGRAPHY

Brandes, Louis Grant, *Yes, Math Can Be Fun.* Portland, Maine: J. Weston Walsh, 1960.

Burns, Marilyn. *The Book of Think.* Boston: Little, Brown & Co., 1976.

Charosh, Mannis. *Mathematical Challenges.* Washington, D.C.: National Council of Teachers of Mathematics, 1965.

Gardner, Martin. *Sixth Book of Mathematical Games from "Scientific American."* New York: Charles Scribner's Sons, 1971.

Greenes, Carole E., Robert E Willcutt, and Mark A. Spikell. *Problem Solving in the Mathematics Laboratory.* Boston: Prindle, Weber & Schmidt, 1972.

Hughes, Barnabas. *Thinking through Problems.* Palo Alto, Calif.: Creative Publications, 1976.

Johnson, Donovan A. *Logic and Reasoning in Mathematics.* St. Louis: Webster Publishing Co., 1963.

Kemeny, John G., J. Lauric Snell, and Gerald L. Thompson. *Finite Mathematics.* Englewood Cliffs, N.J.: Prentice-Hall, 1957.

Salkind, Charles T. *The Contest Problem Book II.* Washington, D.C.: Mathematical Association of America, 1966.

Schaaf, William L. *A Bibliography of Recreational Mathematics,* vol. 4. Reston, Va.: National Council of Teachers of Mathematics, 1978.

———. *The High School Mathematics Library,* vol. 6. Reston, Va.: National Council of Teachers of Mathematics, 1976.

Seymour, Dale, and Margaret Shedd. *Finite Differences.* Palo Alto, Calif.: Creative Publications, 1973.

6

A Senior High School Problem-solving Lesson

Beth M. Schlesinger

THIS paper describes a problem-solving session based on Martin Gardner's "Social Security Number Problem." I have presented this lesson to several classes of students in grades 9–12 with very good results.

I believe that problem solving is an art rather than a skill and that, like other arts, it should be encouraged and disciplined rather than taught as a specific set of techniques to be applied in a rote manner. However, students must be aware of basic techniques in order to use them.

This particular problem is one that can really involve the students. It not only requires insight into the properties of numbers, deductive reasoning, and indirect reasoning but also incorporates a review of all major divisibility tests and relates to lessons on factoring, least common multiple, and greatest common factor. It is therefore a good lesson to present early in the school year.

The problem-solving strategies involved include guessing, breaking a problem into smaller parts, organizing data logically, and using a diagram. Teaching strategies include directed questioning interspersed with minilectures, a modified Socratic method.

The students will need pencil and scratch paper to aid their thought processes.

Problem-solving Session, Part 1

The following dialogue is merely a sample one, of course. At certain points, key responses from the students are required. These can be encouraged by the teacher if they do not occur spontaneously. There are places in the lesson where the students will need time to do calculations or simply to think.

Teacher: Today we are going to spend the class hour working together to solve an interesting and enjoyable problem. It appeared in Martin Gardner's "Mathematical Games" section of *Scientific American* magazine (Gardner 1978). It goes like this:

> Jaime Poniachik of Buenos Aires . . . said he had a friend in the U.S. with a curious social security number: its nine digits include every digit from 1 through 9, and they form a number in which the first two digits (reading from left to right) make a number divisible by 2, the first three digits make a number divisible by 3, and the first four digits make a number divisible by 4 and so on until the entire number is divisible by 9. What is the number?

Teacher: Does everybody understand the problem?

Student: You mean we have to find one long number?

T: Right.

S: How about 123 456 789?

T: Let's see . . . 12 is divisible by 2 . . . 123 is divisible by 3 . . . 1234 is *not* divisible by 4. So your guess breaks down at 4.

S: I know! 123 645 789.

T: If you test it out, you'll see that that number won't work, either. We have seen one approach to solving a problem—the "guess" method. It's a method that students often use; unfortunately, it doesn't work too often. But it's a way of starting to solve a problem. What we are going to do today is attack this problem using logic and our knowledge of how numbers work. Do you think a diagram would help?

S: I know. We need a series of blanks.

T: Good! And let's number them. [*Writes on the chalkboard.* See fig. 6.1.]

___	___	___	___	___	___	___	___	___
1	2	3	4	5	6	7	8	9

Fig. 6.1

T: Now, each of the digits 1 through 9 goes into exactly one space.

S: Is there more than one answer?

T: I'm not sure. I know there is at least one answer. Later on we'll try to decide if the solution is unique. To start with, one number can be placed in its correct space right away. Which one? [*Pause.*]

S: The 5 has to go in space 5.

T: Excellent! Why is this so?

S: We know that if 5 divides a number evenly, the number must end in either 5 or 0.

T: And we're not using 0. So now we have this. [*Writes.* See fig. 6.2.]

___	___	___	___	5	___	___	___	___
1	2	3	4	5	6	7	8	9

Fig. 6.2

T: Let's review our other divisibility tests. A number is divisible by 2 if . . .
S: . . . it ends in 2, 4, 6, 8, or 0.
T: Correct. A number is divisible by 3 if . . .
S: . . . the sum of its digits is divisible by 3.
T: Right. A number is divisible by 4 if . . .
S: . . . the number that the last two digits make is divisible by 4.
T: A number is divisible by 6 if . . .
S: . . . it's divisible by both 2 and 3.
T: A number is divisible by 7 if . . . [*Silence.*] We'll discuss this test later. Does anybody know the test for 8? [*Pause.*] No? A number is divisible by 8 if the number named by its last three digits is divisible by 8. For example, 1824 is divisible by 8 because 824 is divisible by 8. Now a number is divisible by 9 if . . .
S: . . . the sum of its digits is divisible by 9.
T: Does anybody have any other thoughts so far? Well, then, let me ask you, which choices of digits do we have for spaces 2, 4, 6, and 8?
S: Each of these will have to be an even number.
T: Can we tell which even number?
S: No.
T: So we have this situation. [*Writes.* See fig. 6.3.]

___	___	___	___	5	___	___	___	___
1	2	3	4	5	6	7	8	9
	2		2		2		2	
	4		4		4		4	
	6		6		6		6	
	8		8		8		8	

Fig. 6.3

T: What else do we notice?
S: We have four spaces where even numbers have to go, and four even numbers. So that uses up the even numbers.
T: Where will the odd numbers go?
S: They will go in the odd-numbered spaces. But 5 is used up already.
T: Right! So now we have this. [See fig. 6.4.]

___	___	___	___	5	___	___	___	___
1	2	3	4	5	6	7	8	9
1	2	1	2		2	1	2	1
3	4	3	4		4	3	4	3
7	6	7	6		6	7	6	7
9	8	9	8		8	9	8	9

Fig. 6.4

T: What next? [*Silence.*] How can we use a divisibility test for 4?

S: The digits in spaces 3 and 4 must form a number divisible by 4.

T: How many such possible numbers are there? Can you name them?

S: Four possibilities for space 3 times four for space 4 equals sixteen possible numbers. They are 12, 14, 16, 18, 32, 34, 36, 38, 72, 74, 76, 78, 92, 94, 96, and 98.

T: Which ones are divisible by 4?

S: 12, 16, 32, 36, 72, 76, 92, and 96. Look! There will be a 2 or a 6 in space 4.

T: Good! [*Crosses out the 4 and 8 under space 4.*] Now, let's talk about using our divisibility test for 3. What do you know about the first block of three digits, that is, spaces 1 through 3?

S: The sum of these three digits must be divisible by 3.

T: All right. Now what about spaces 1 through 6?

S: The sum of these six digits must be divisible by 3 also.

T: Can you conclude anything about spaces 4 through 6 alone? Remember, if a sum is divisible by 3, each addend must be divisible by 3. That is, $3x + 3y$ is divisible by 3, but $3x + y$ is not always divisible by 3.

S: I see. The sum of spaces 4, 5, and 6 must be divisible by 3 also.

S: And the sum of the digits in spaces 7, 8, and 9 must be divisible by 3.

T: Right! Now let's focus on spaces 4, 5, and 6. How many possible choices do we have here? Can you name these numbers?

S: $2 \times 1 \times 4 = 8$ numbers. They are 252, 254, 256, 258, 652, 654, 656, and 658.

T: Can we eliminate any of these?

S: 252 and 656 because some digits are used more than once.

T: Good. Which of the remaining numbers are divisible by 3? [*Time for calculation.*]

S: Only 258 and 654.

T: So now we have this. [See fig. 6.5.]

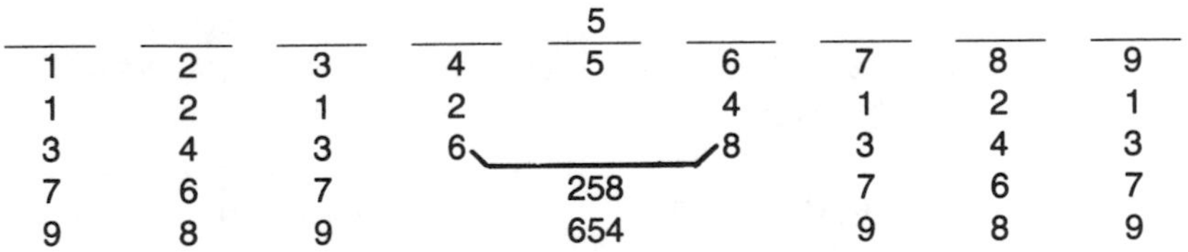

				5				
1	2	3	4	5	6	7	8	9
1	2	1	2		4	1	2	1
3	4	3	6		8	3	4	3
7	6	7		258		7	6	7
9	8	9		654		9	8	9

Fig. 6.5

T: We've gone through all the divisibility tests up through the one for 6. Let's skip the one for 7 for now. But what about 8? Looking at spaces 6, 7, and 8, we'll have to check out thirty-two possibilities ($2 \times 4 \times 4 = 32$). But we'll find shortcuts. It's important to organize our attack on a problem.

Let's list the possibilities and cross out repetitions right away. [See fig. 6.6.]

412	432	472	492	812	832	872	892
~~414~~	~~434~~	~~474~~	~~494~~	814	834	874	894
416	436	476	496	816	836	876	896
418	438	478	498	~~818~~	~~838~~	~~878~~	~~898~~

Fig. 6.6

T: Now let's look at the first column . . . 416 is divisible by 8, and we can skip checking 412 or 418. Why?

S: Because 424 will be the next higher number divisible by 8, and 408, the next lower.

T: Exactly . . . 432 works. And we can skip the others in that column because 440 will be the next multiple of 8. Now go through the other columns to find out which numbers are divisible by 8. [*Time for calculation.*]

S: 416, 432, 472, 496, 816, 832, 872, and 896.

T: Now what about spaces 7, 8, and 9 taken together?

S: They must name a number divisible by 3.

T: We need to take 16, 32, 72, or 96 for spaces 7 and 8 and combine these with 1, 3, 7, or 9 from space 9 to see which ones give us numbers divisible by 3. And remember we can't repeat digits. Let's make another list. [See fig. 6.7.]

~~161~~	321	721	961
163	~~323~~	723	963
167	327	~~727~~	967
169	329	729	~~969~~

Fig. 6.7

T: Now apply your test for 3 and tell us which ones work. [*Calculation time.*]

S: The only possibilities for spaces 7, 8, and 9 are 321, 327, 723, 729, or 963.

T: Do we all agree to this? Good. We're getting closer to our solution.

S: It's like solving a mystery.

T: Yes. Now we have this. [See fig. 6.8.]

S: The 2 and the 6 will be used in spaces 4 and 8, and so that leaves only a 4 or an 8 for space 2.

T: Right. [*Crosses out the 2 and 6 under space 2.*] Now what about the block consisting of spaces 1, 2, and 3? Remember, this number must be divisible by 3. How many possibilities are there, and what are they?

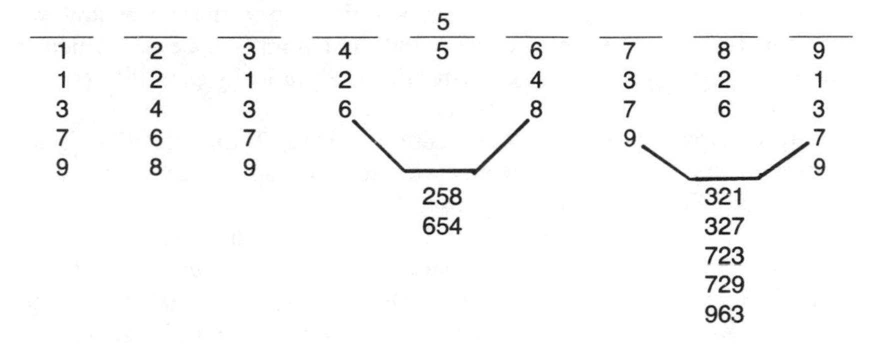

Fig. 6.8

S: $4 \times 2 \times 4 = 32$ possibilities. [*Helps make the list in fig. 6.9.*]

~~141~~	~~181~~	341	381	741	781	941	981
143	183	~~343~~	~~383~~	743	783	943	983
147	187	347	387	~~747~~	~~787~~	947	987
149	189	349	389	749	789	~~949~~	~~989~~

Fig. 6.9

T: Which of these are divisible by 3? [*Calculation time.*]

S: I get ten possible numbers: 147, 183, 189, 381, 387, 741, 783, 789, 981, and 987.

T: How many others got this answer? Good. Now we have this. [See fig. 6.10.]

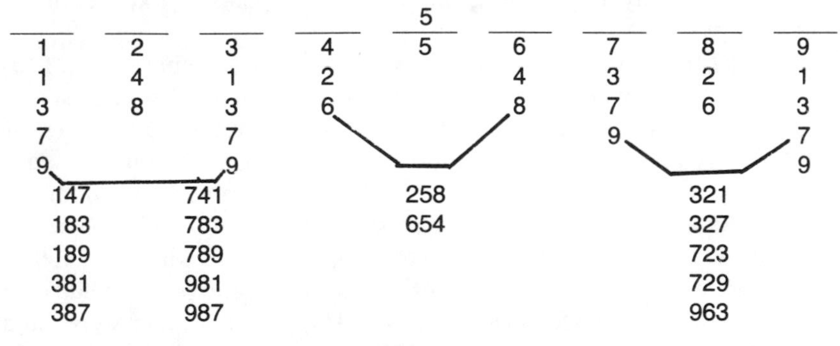

Fig. 6.10

T: The only two divisibility tests we haven't used are the ones for 7 and 9. Will the test for 9 be helpful? [*Pause.*]

S: No. Because the sum of the numbers 1 through 9, no matter how we arrange them, will always be the same. The sum will be 45, which is divisible by 9. So any arrangement will automatically pass the test for divisibility by 9.

T: Correct! Now I'll show you the divisibility test for 7 (Linden 1974). Take a number, say 1 234 567. Starting at the right, group the number by twos: 1 23 45 67. The *leftmost group* may have one or two digits. Take the number in this group and double it: $2 \times 1 = 2$. Add the next group to the right: $2 + 23 = 25$. Double this *sum* and add the next group to the right: $2 \times 25 = 50$, and $50 + 45 = 95$. Keep on in this pattern: $2 \times 95 = 190$, and $190 + 67 = 257$. If your final result is more than two digits long, divide it as at the start and repeat the procedure: 2 57. Now $2 \times 2 = 4$, and $4 + 57 = 61$. You will end up with a one- or two-digit number. If this number is divisible by 7, your original number is also divisible by 7. Since 61 is not divisible by 7, our original number, 1 234 567, is not divisible by 7. Let's try another example: 8 641 969. Divide it up: 8 64 19 69. Then $2 \times 8 = 16$, and $16 + 64 = 80$. $80 \times 2 = 160$, and $160 + 19 = 179$. $179 \times 2 = 358$, and $358 + 69 = 427$. Divide up 427: 4 27. Then, $2 \times 4 = 8$, and $8 + 27 = 35$. Thirty-five is divisible by 7, and so is 8 641 969.

T: Well, we've gone far enough in class today. I want you to take this problem home and work on it some more. I will give you two hints: First, try to decide whether a 4 or an 8 goes in space 2. Second, use your new divisibility test for 7. Organize your work so you can explain your thought processes and then make deductions. We're fairly close to a solution even though only one number has been placed! Good luck!

Problem-solving Session, Part 2

Teacher: Does anybody have a solution? Three people? Excellent! Mia, let's hear yours.

Mia: First I tried to decide whether a 4 or an 8 would go in space 2. I first assumed that a 4 would work. So I had 147 or 741 in the first three spaces. Since the 4 was used up, 258 had to go in the next three spaces with a 9 in space 7. (I couldn't use a 2 twice, and all the other combinations for spaces 7, 8, and 9 used a 2 in space 8.) So I had two seven-digit numbers: 1 472 589 and 7 412 589. When I used the divisibility test for 7, neither passed. So I concluded that an 8 had to go in space 2. This meant that the 4 had to go in space 6, and so 654 went in spaces 4, 5, and 6. I then had eight possible combinations for spaces 1, 2, and 3; spaces 4, 5, and 6 were filled; and I had only two possible numbers for space 7. (I couldn't use 963 for the last three spaces, since the 6 was already used.) I stopped with space 7 because I planned to use the divisibility test for 7. I made another diagram. [*Writes on the chalkboard.* See fig. 6.11.]

		654	
1, 2, 3		4, 5, 6	7
183	~~783~~		3
189	789		7
381	981		
~~387~~	987		

Fig. 6.11

I crossed out 387 and 783 right away because I knew I needed either a 3 or a 7 for space 7. Then I listed all possible seven-digit numbers, crossing out the ones with repeated digits [fig. 6.12].

~~1836543~~	~~3816543~~	9816543
1836547	3816547	9816547
1896543	7896543	9876543
1896547	~~7896547~~	~~9876547~~

Fig. 6.12.

Out of sixteen original possibilities, I had only eight to check. I used the divisibility test for 7 on these eight numbers. Only one of them was divisible by 7: 3816547. So I concluded that the number must be 381 654 729. Since we are looking for a social security number,it would be written 381-65-4729.

T; That's exactly right! [*Applause.*] Did you other students get the same answer? Good! Notice that Mia used indirect reasoning to see which number went in space 2, the 4 or the 8. She assumed that the 4 would work and made deductions until she met a contradiction. It's the same as with an indirect proof.

T: Now, Mia, do you think that this solution in unique?

S: It has to be. It's the only combination that met all the requirements.

T: To close this problem-solving session, we should be proud of ourselves—we solved a difficult problem. We went about it step by step, used basic facts about numbers and divisibility tests, and used deductive and indirect reasoning. Our diagram was extremely helpful. I hope you've enjoyed this lesson.

Follow Up

To double-check the uniqueness of the answer to this problem, Steve Robbins, a twelfth grader at Gompers Secondary School, wrote the following computer program in BASIC-PLUS. (The computer on which the problem was run is a Digital Equipment Corporation computer, model PDP-11/60, using RSTS, version 7.0.)

```
10 DIM DIG% (9%)
20 FOR FIRST = 1. TO 9.
25 DIG% (FIRST) = -1%
30 Z = FNG (FIRST, 1%, 0%)
40 IF Z<>-1. THEN GOTO 200
45 DIG% (FIRST) = 0%
50 NEXT FIRST
60 PRINT 'No such number'
70 STOP
200 PRINT 'The number was ';NUM1$(Z)
210 FOR X% = 2% TO 9%\ ZZZ = INT (Z/10^(9-X%))
220 PRINT NUM1$ (ZZZ); '/';NUM1$ (X%); ' = ';NUM1$ (ZZZ/(X%*1.))
230 NEXT X%
290 GOTO 32767
300 DEF* FNG(SO.FAR, LNG%, X%)
310 FOR X% = 1% TO 9%
320 NEWLNG% = LNG% + 1%
330 GOTO 450 IF DIG% (X%)
340 Z = SO.FAR*10. + X%
350 GOTO 450 IF INT (Z/NEWLNG%) *NEWLNG%<>Z
355 IF NEWLNG% = 9% THEN GOTO 380
360 DIG% (X%) = - 1%\ Z= FNG (Z, LNG% + 1%, 0%)
370 IF Z = - 1. THEN DIG% (X%) = 0%\GOTO 450
380 FNG = Z\GOTO 500
450 NEXT X%\FNG = -1.
500 FNEND
32767  PRINT TIME (1%)\END
```

RUN SSN

```
The number was 381654729
38 divided by 2 = 19
381 divided by 3 = 127
3816 divided by 4 = 954
38165 divided by 5 = 7633
381654 divided by 6 = 63609
3816547 divided by 7 = 545221
38165472 divided by 8 = 4770684
381654729 divided by 9 = 42406081
Ready
```

REFERENCES

Gardner, Martin. "Mathematical Games." *Scientific American,* December 1978, p. 23.

Linden, Andrew, "Divisibility by 7." *Mathematics Teaching* [England], no. 68, September 1974, p. 143.

7

Identification and Analysis of Specific Problem-solving Strategies

Hunter Ballew

E*DUCATORS should give priority to the identification and analysis of specific problem-solving strategies.* That statement comes from the discussion of Recommendation 1 of *An Agenda for Action,* published by the National Council of Teachers of Mathematics in 1980. In this article, I propose to identify, analyze, and illustrate two closely related problem-solving strategies. The first strategy is to reduce a complex problem to a series of simpler problems, and the second strategy is to use a numerical analogy to help develop a plan of attack that boosts the confidence of the problem solver. The type of analogy referred to here usually involves, say, substituting a smaller number for a larger one or perhaps a whole number for a fraction to aid the reasoning process of the problem solver. It should be noted that these strategies are not restricted to conventional word problems but are also frequently useful in solving nonroutine problem situations.

Reducing a Complex Problem to a Series of Simpler Problems

This strategy can be clearly illustrated with Krypto. Krypto is a commercially made learning activity that produces a rich pool of problems of varying difficulty. Some problems arising from Krypto are simple, some are quite challenging, some are even impossible, and some may *seem* either simple or complex, depending on the learner's mental set at the time the problem is encountered.

Briefly, each participant is dealt a hand of five cards, each card with a single numeral from 1 through 25 on it. After each participant receives a hand, a single card from the remaining cards is then dealt for all participants to share as an object card. The goal is to use each of the five numbers in a hand to form the left side of an equation. The right side of the equation must be the number on the shared object card. In forming the left side of the equation, participants may choose to use any one of the four operations of addition, subtraction, multiplication, and division, or any combination of these four operations. Each of the five cards of a hand must be used once and only once on the left side of the equation, as in the following example:

Hand: 3, 8, 8, 25, 2

Shared object card: 15

Solution: $(8 + 25) - 3\,(8 - 2) = 15$

Many hands can be solved by a brief inspection. The purpose of a cognitive strategy is to provide a systematic approach if this initial inspection fails to solve the problem. A cognitive strategy helps students find some course of action to take when they do not know exactly how to solve a particular problem. The course of action *may* or *may not* lead to the solution, and this feature distinguishes a cognitive strategy from an algorithm. Consider another example:

Hand: 3, 7, 14, 12, 4

Shared object card: 24

Some students might find a solution after a brief consideration of the numbers. Many are not among that fortunate group, however, and need to attack the problem in a systematic way by breaking it down into a series of subproblems. This can be accomplished by simply omitting one of the cards from the hand. This reduces the hand to four cards, and this reduced hand often will be easier to work with than five cards. Furthermore, this introduces some new "object" cards for the right side of the equation, creating possibilities that *may* be easier to investigate than the original problem. In the sample hand, suppose we arbitrarily decided to eliminate the 4:

New Hand: 3, 7, 14, 12

New possible object cards are the following:

a) 20, because 20 could then be added to the eliminated 4 to produce the original object card of 24

b) 28, because the eliminated 4 could then be subtracted from 28 to produce 24

c) 6, because 6 can then be multiplied by 4 to produce 24

d) 96, because 96 divided by 4 is 24

By examining the new four-card hand and combining the numbers in different ways, we can determine that 6 (possibility *c* above) can be obtained: 12 divided by 3 is 4, and 14 divided by 7 is 2, and the sum of these quotients is 6. Now the complete solution to the original problem can be written:

$$\left(\frac{12}{3} + \frac{14}{7}\right) \times 4 = 24$$

If the arbitrary elimination of 4 from the original hand does not lead to a solution, another card could be eliminated, leaving a different four-card hand with which to work. For example, the 12 might be eliminated, leaving 3, 7, 4, and 14. If we could now produce a 2 from these four numbers, we could then multiply the 2 by the previously eliminated 12, thus getting the required 24. A 2 can be obtained from these four numbers like this:

$$\frac{14}{7} \times (4 - 3) = 2$$

All that remains is to write the solution to the original problem.

If none of the four-card hands produces a solution, we might then systematically reduce the four-card hands to a series of three-card hands. And these could subsequently be reduced to two-card hands if need be.

The power of this particular cognitive strategy is shown by the fact that this systematic reduction of a complex problem either will eventually produce a solution or will prove that no solution exists. For example, a hand consisting of 3, 10, 17, 17, and 25, with a shared object card of 24, can be analyzed into all possible subhands, and it can be shown that no solution exists.

Nonroutine problems assume greater educational value if they are not treated as isolated examples. For instance, if Krypto were used in the classroom simply as a game, it would provide a lot of entertainment but it would lose its potential as a tool for the development of mathematical power. Playing a few hands makes it quite obvious that Krypto provides an excellent context for computational practice and equation construction. What is not quite so obvious is that Krypto can also be used to help students develop a problem-solving strategy useful in other mathematical situations.

The algorithms we so quickly and easily use today for addition, subtraction, multiplication, and division were probably developed painfully over a long period of time by people who unlocked the secret of changing a single problem into a series of problems. For example, if we are asked to add two numbers such as 235 and 462, we know that by following place-value rules, we can change this problem into the following series of problems: 5 + 2, 3 + 6, and 2 + 4. At one time in the development of our civilization, 235 + 462 would doubtless have been a formidable problem to almost anyone. Today, however, any child who understands place value and knows the basic addition facts can find this sum quickly and easily.

Unit fractions provide a context for further development and use of the strategy of breaking a single problem into a series of problems. A unit fraction is a simple fraction having 1 as the numerator. Let us admit right away that unit fractions do not constitute a classification *ordinarily* useful in the study of mathematics. We usually classify fractions as simple or complex, proper or improper, or common or decimal. In planning mathematics instruction, however, we find that a knowledge of unit fractions is useful for three reasons:

1. Some study of cultural background can be built into mathematics lessons because, with the exception of 2/3, unit fractions were the only kind of fractions used by the ancient Egyptians.
2. Work with unit fractions provides a different context for practice in adding and subtracting fractions and in finding equivalent fractions.
3. Nonroutine problems can be posed using unit fractions, and these can be solved by using the strategy of breaking a single problem into a series of problems.

The last reason is the focus of this article. Our nonroutine problem posed with unit fractions is this:

> **Write a given fraction as the sum of two or more positive unit fractions, no two of which are equal.**

With some unit fractions, this problem is solved by inspection. For example, 3/4 may be written as 1/2 + 1/4, and 5/6 may be written as 1/2 + 1/3. Many fractions provide considerably more difficult problems, however. Inspection does not provide a ready solution for 4/7. A solution of 1/7 + 1/7 + 1/7 + 1/7 cannot be accepted because it violates the given condition that all the unit fractions must be different. (Some students propose 3/7 + 1/7 as a solution. These students have to be reminded that 3/7 is not a unit fraction.)

In the example of 4/7, a new strategy is needed. Left to their own devices after false starts similar to the ones mentioned earlier, some students in a class will usually discover the strategy of changing the given fraction to equivalent fractions. When this happens, this strategy should be discussed with the entire class. Students who develop this strategy can explain their methods and results to the class. A solution for 4/7 is

$$\frac{4}{7}=\frac{8}{14}=\frac{7}{14}+\frac{1}{14}=\frac{1}{2}+\frac{1}{14}.$$

The strategy of changing to equivalent fractions leads to quick success with many fractions. With many other fractions, however, the numbers grow large rapidly, and solutions are easily overlooked. For example, if 9/11 is the given fraction, an approach to the problem would be to write the following:

$$\frac{9}{11}=\frac{18}{22}=\frac{27}{33}=\frac{36}{44}.$$

The first three of these equivalent fractions yield no solution. The fourth one, 36/44, does yield a solution, but it could easily be overlooked. A systematic and careful approach is necessary to spot the solution. In order to develop a systematic approach that will prevent overlooking solutions, mathematical thinking that leads to two observations is necessary. That is:

1. The given fraction must be written as a series of fractions so that the sum of the numerators is equal to the given numerator.
2. Each of the numerators in the selected series of fractions must be a factor of the denominator so that a unit fraction can be obtained.

A systematic way to fulfill both these conditions simultaneously is to write all the factors of the denominator except the denominator itself and then see if any combination of these factors will add up to the numerator. For example, write 9/11 as 27/33. The factors of 33 that are of possible use are 1, 3, and 11. No combination of these three factors will add up to 27, and so we must consider another equivalent fraction. Write 9/11 as 36/44. The factors of 44 that are of possible use are 1, 2, 4, 11, and 22. Trying various combinations of these factors, we find that $22 + 11 + 2 + 1 = 36$. Therefore a solution is

$$\frac{9}{11} = \frac{36}{44} = \frac{22}{44} + \frac{11}{44} + \frac{2}{44} + \frac{1}{44} = \frac{1}{2} + \frac{1}{4} + \frac{1}{22} + \frac{1}{44}.$$

The equivalent fraction strategy works well in many cases, but it tends to lose its efficiency with larger numbers. For example, if the equivalent fraction strategy is applied to 13/19, no solution can be found until the equivalent fraction 78/114 is obtained. Therefore, the application of the strategy of changing a single problem to a series of problems might seem appropriate. We could reason that the given fraction, 13/19, is more than 1/2, and then by taking the 1/2 from 13/19, we would have part of the answer and a new problem to solve, namely—

$$\frac{13}{19} - \frac{1}{2} = \frac{26 - 19}{38} = \frac{7}{38}.$$

Part of our answer is 1/2, and the new problem is to write 7/38 as the sum of unit fractions. Continuing this process has the effect of changing a single problem into a series of problems. We would now subtract a unit fraction from 7/38, and the difference becomes yet another problem in the series:

$$\frac{7}{38} - \frac{1}{6} = \frac{42 - 38}{228} = \frac{4}{228} = \frac{1}{57}$$

The solution to the original problem is now seen to be

$$\frac{13}{19} = \frac{1}{2} + \frac{1}{6} + \frac{1}{57}.$$

The first objection students usually make to this is, "How did you know to subtract the 1/6?" The answer is, "You don't know until you try." Frequently it may be advantageous to subtract the largest possible unit fraction, which can be determined by some simple arithmetic. From a psychological point of view, however, inhibitions to thinking are lessened when students realize it is not absolutely necessary to always subtract the largest possible unit fraction. An interesting classroom observation is that more than one correct solution is frequently found. For example, if we had subtracted 1/8 instead of 1/6, we would have obtained the following:

$$\frac{7}{38} - \frac{1}{8} = \frac{56 - 38}{304} = \frac{18}{304} = \frac{9}{152}$$

In this example, our partial solution is 1/2 + 1/8, and the new problem is 9/152. The solution, which is different from the earlier one, is

$$\frac{13}{19} = \frac{1}{2} + \frac{1}{8} + \frac{1}{17} + \frac{1}{2584}.$$

A lot of good arithmetic practice can be gained in checking the solution—an added benefit derived from studying unit fractions.

There is an added protection for the student trying to find an acceptable fraction to subtract from a given fraction. If too large a fraction is subtracted, a negative result is obtained, which is a signal to go back and select a smaller fraction. For example, suppose we had subtracted 1/5 from 7/38:

$$\frac{7}{38} - \frac{1}{5} = \frac{35 - 38}{190} = \frac{-3}{190}$$

Since negative unit fractions are not allowed by the conditions set forth in the problem, this is a signal to retrace our steps and to subtract a smaller fraction.

Developing and Using Analogies in Solving Problems

In some situations an analogy might be more helpful than a series of subproblems in pointing the way toward a plausible solution. For example, consider the following problem dealing with the rate of gasoline consumption:

Maria drove 595 kilometers on 46.5 liters of gasoline. How many kilometers per liter of gasoline did she get?

Students frequently become confused when attempting to solve problems of this type. They may be fairly certain that they should either multiply or divide, but they lack the confidence to select the correct operation. An analogy can often supply that missing confidence. Instruction can be designed to help students develop the ability to create analogies by rewriting given problems and substituting numbers they can cope with more easily. In the problem above, the question has to do with kilometers *per liter*—a clue that a substitution of 1 liter for the 46.5 liters in the original problem might provide a helpful analogy. A substitution could also be made for the other number in the problem if the student finds it helpful. The point is that different students will create different analogies, and the process can be refined and shortened as each student gains confidence in interpreting and solving problems. A student who is just beginning to develop this problem-solving strategy might create the following analogy to the given illustration:

> **Maria drove 50 kilometers on 1 liter of gasoline. How many kilometers per liter of gasoline did she get?**

The answer to this analogous problem is obviously 50 kilometers per liter. The answer is so obvious, in fact, that the student would probably not be aware *from this analogy alone* of just what mathematical process was involved. The student using this analogy must now be instructed to take advantage of the confidence instilled by the obviously correct answer and create and solve an analogy slightly more difficult. For example:

> **Maria drove 50 kilometers on 2 liters of gasoline. How many kilometers per liter of gasoline did she get?**

The answer is 25, and most students could accept this answer with confidence. Students using this analogy should subsequently be instructed to ask themselves, "What mathematical process did I use to get an answer of 25 from the given 50 kilometers and 2 liters?" Some students could be expected to apply the same process with confidence to the original problem. Others might need to approach the original problem more slowly through some additional intermediate analogies. These students need to be instructed to ask themselves questions in which the number of kilometers and the number of liters vary. A helpful strategy is for students to organize the answers to their own series of questions in tables so that patterns can be discerned and trends checked for consistency.

Distance	Amount of Gasoline	Kilometers/ Liter
50	1	50
50	2	25
100	2	50
200	2	100
500	2	250
595	2	297.5
595	3	198.3
595	5	119.0
595	10	59.5
595	46.5	12.8

This much detail will be unnecessary for many students. However, if the students who usually have great difficulty with this type of problem will carefully work through such a sequence, with instruction as needed from the teacher or from fellow students, they will have confidence that the bottom line of the table provides the correct answer to the original problem.

Another type of problem appropriate to the analogy strategy is the conversion problem involving the customary and the metric systems of measurement. Such problems are *not* recommended as a means of *introducing* the metric system to youngsters, but they do provide excellent applications of the very important mathematical concept of ratio. Examples such as the following are appropriate for middle school and high school students:

If 0.62 mile equals 1 kilometer, how many kilometers are in 50 miles?

Many students are quite willing to attack this problem by multiplying 0.62 times 50. Many even check their multiplication and become quite satisfied that 31 kilometers is the answer. A quick application of even a modest amount of estimation skill would reveal 31 to be an entirely unreasonable answer. Students who get 31 kilometers should be made to feel that it is perfectly all right to obtain 31 *if* they will treat it as a tentative answer. However, they must go further and check the reasonableness of the answer. These students should be instructed to ask themselves, "Will there be more or fewer than 50 kilometers in 50 miles?" As they ponder this question, they see that their initial answer of 31 is suspect and realize that they must rethink their approach to the problem. What they need is an approach that will inspire a high degree of confidence in the forthcoming solution. A number simplification can provide that needed confidence. One helpful analogy that could be created by a student is the following:

If 0.5 mile equals 1 kilometer, how many kilometers are in 1 mile?

The answer, as seen by most students with confidence, is 2. As in the previous illustration, substitutions of other numbers can be made and the results of these simplifications tabulated.

Miles	Kilometers
0.5	1
1	2
4	8
10	20
50	100

This table is based, of course, on the approximate relationship of 0.5 mile equaling 1 kilometer. The answer for 50 miles provided by the analogy is not exact, but the process for obtaining the answer *is* correct. Displayed in a table, the data help students conclude that the correct process for converting miles to kilometers is to divide the total number of miles by the number of miles in one kilometer. In this analogy, 0.5 is divided into 50 to produce an answer of 100. Some students might now return to the original problem and divide 0.62 into 50 to obtain the correct answer. Other students might feel more comfortable in constructing a new table based on the given ratio:

Miles	Kilometers
0.62	1
1	$1 \div 0.62$
2	$2 \div 0.62$
10	$10 \div 0.62$
50	$50 \div 0.62$

The purpose of an analogy is to guide the students' thinking into correct channels. After solving the analogous situation, the student can apply the developed procedures to the original problem. The analogy strategy can be applied to many other types of problems, including the following:

1. More complicated conversions, such as those between Celsius and Fahrenheit temperatures, provided it is known or given in the problem that 0°C= 32°F and 100°C = 212°F
2. Conversions within systems, such as changing square centimeters to square meters
3. Unit pricing
4. Distance-rate-time problems
5. Principal-rate-interest problems
6. Percentage problems

Summary

Many problems of the type used as illustrations in this article can be solved efficiently by more mechanical means. For example, conversion problems can be solved quite readily by proportions. Although the efficiency of these more mechanical means must be acknowledged, caution should be exercised against an overdependence on such methods. First, mechanical means sometimes have a way of becoming substitutes for real mathematical thinking. Second, strategies such as breaking a problem into a series of subproblems and creating analogies can be used before students develop algebraic procedures. Third, mechanical procedures are easily forgotten and easily misapplied or confused with other procedures. Problem-solving strategies are especially helpful when students are confronted with several different kinds of problems at one time and are not certain what procedure to follow. Students who feel comfortable with several well-developed strategies have a good chance of finding a place to start on a problem. In short, strategies help students *find* what to do when they do not already *know* what to do.

Recommendation 2

THE CONCEPT OF BASIC SKILLS IN MATHEMATICS MUST ENCOMPASS MORE THAN COMPUTATIONAL FACILITY

8

The Statistical Survey: a Class Project

Murray H. Siegel

THE decade of the eighties is an age of awareness. With television, videotape, microminiaturization, and satellite technology, little occurs in the world that is not brought to the attention of most citizens. This state of awareness has significant implications for the mathematics classroom of the eighties. More than ever before, students, parents, and taxpayers are asking, "Why is this taught? Why is it important?" The mathematics curriculum must include exercises that demonstrate the relevance of the mathematical subject matter. Ideally, such exercises should be understandable to students yet obviously relative to real-world situations. They should also require a number of mathematical skills and should, furthermore, include skills from other academic areas. The inclusion of other disciplines in the exercise would make it appear to be more realistic and less of a "textbook type" of problem.

During the past six years I have developed such an exercise. My experience indicates that it could be adapted to grades 5–8. The exercise is the statistical survey.

For many people, statistics is a subject that is not examined until college. Yet many elementary school curricula include topics in statistics and probability at all grade levels. At this point, you may ask just what mathematics must be understood by a student to complete the statistical survey. The student should be able to compute using fractions and decimals and should also be able to compute the mean of a set of data. The latter may be accomplished by adding up the measurements and dividing by the number of measurements or by using a frequency distribution. The student should be able to construct a bar graph from a given set of data. The student should understand the concept of probability as a fraction with the numerator being the number of times a particular event occurred and the denominator being the sample size. Two optional topics that are not vital but do lend breadth to the results are percent and the computation of standard deviation. Standard deviation may seem beyond the ability of the typical middle-grade student;

yet my personal experience has shown that this computation can be accomplished by gifted students as early as fifth grade, using calculators for the individual computational steps.

The only purpose for a statistical survey is to answer a question (or questions) that is important to some person or organization. A manufacturer might have a marketing study performed to answer such questions as, "Will the consumer purchase my new product?" "Will a new package design enhance the public's perception of my product?" "Is our recent advertising campaign having positive effects on sales?" "How does the consumer view my product compared to the product of my competitors?" A politician could have a survey taken to answer, "How do my constituents view my record?" "How many voters recognize my name?" "What portion of the elecorate would vote for me?"

Statistical surveys are also compiled to rate radio and television stations and shows and to acquire insight into the public's mood on issues and current events. Thus a statistical survey begins with a question whose answer will offer a solution to a problem, or at least point the way to an appropriate problem-solving process.

To begin the project, each student must devise a question that is important to some person or organization. Also, the answers obtained must be in the form of numerical data. If the data are not in numerical form, then mathematical computations cannot be performed. Examples of meaningful questions that have been used in projects done by my students include the following:

- What is the public attitude toward the recognition of the People's Republic of China?
- How do shoppers rate the prices at a particular discount store?
- How many telephones are installed in the average residence?
- How often does the average person eat at a particular fast-food restaurant?
- How do viewers rate a particular local news program?
- How do listeners evaluate a particular disc jockey?
- How fast do automobiles travel through residential neighborhoods?

How are students to develop meaningful questions? What I have done in my classes is discuss the use of surveys in situations relevant to the students' world. Usually we discuss radio ratings and TV Nielsen ratings. If an election is upcoming, we discuss straw polls that have been held to predict the election results. Once the notion of a survey is established, I relate to the class some of the questions that have been used by students in the past. Then each student is required to write out a question that seems important, indicating why and to whom the question is of concern. Each student's output is evaluated in a one-to-one conference with the teacher. When a student has difficulty getting started, the teacher should draw out the child's interests and try to relate them to potential survey questions.

Once the topic for the question is determined, the wording must allow for converting results to numerical answers. There are three basic methods of constructing questions that will provide numerical data. The obvious method is to ask a question that requires a numerical answer (e.g., *how many times, how long since,* or a specific measurement such as length, speed, or weight). Another manner of questioning is to ask for a numerical rating (e.g., "On a scale of 1 to 10, with 10 being the highest, please rate. . ."). The third method uses a statement for which respondents indicate their opinion (*strongly agree, agree, no opinion, disagree,* or *strongly disagree*). A number is then assigned to each response (strongly agree is equivalent to 5; agree, to 4; etc.).

It will be determined that sometimes an introductory question is necessary. That is, if I am going to ask a person to evaluate Channel 14 movies, I must first ask whether that person watches these movies. If a preliminary question must be asked, then a count of the number of people who answer in the negative should be maintained and included in the survey report.

If the question involves a comparison between two stores, two restaurants, or two programs, then two introductory questions are needed to ensure that respondents are familiar with the two things being compared.

Once the students have determined *what* question or questions are to be asked, it is vital to establish *how* each question will be asked. Each student presents his or her question to the class. The class discusses whether the question is grammatically correct and clearly worded. Yes, this is a mathematics class, but proper grammar and clear wording are important, since we want those being interviewed to understand what is being asked. During this discussion, the class also determines if the answers can be converted to numerical data. This discussion will help students determine if their question is worded in the manner most likely to provide the data desired. For example, Joe might ask, "Can you name a U.S. president who was also a five-star general?" or he can ask, "Did you know that Dwight Eisenhower was both a five-star general and president of the United States?" However, if he says instead, "Name a president of the U.S. who was also a five-star general," he will more likely be able to obtain consistently useful answers. The first two phrasings of the question are far less likely to produce specific answers.

Now it is necessary to establish among the students that each subject must be asked precisely the same question. The class should be asked for suggestions to ensure that the questioning is consistent. The most effective way to do this is for the interviewer to write the question on a piece of paper and read it to each person surveyed. Another point to be made in class is that the interviewer should not ask the question of people who have heard the answers given by other respondents because their opinion mught be influenced by what they have heard.

For a question to be important, it must be relevant to a large population. Since we cannot survey all the adults in Little Rock, all those who eat school lunches in Cook County, or all Americans who buy jeans, we must use a *representative sample.* A discussion of sampling is a valuable social studies lesson as well as a good mathematical topic. Information on radio station ratings can be obtained from a local radio station or a rating service (if there is one in your town). How many people are sampled? How are they chosen? What were the results of the most recent ratings? The teacher might ask the class if they know anyone who was ever questioned about the radio stations to which they listen. The Nielsen ratings, which determine which TV shows live and which ones die, is based on a sample of about 1200 homes. Sampling techniques are used in cooking or in product testing (e.g., *Consumer Reports*).

The first task in selecting a sample is to identify the population. Age, sex, and geographical placement may be important elements. If the question concerns cable television, then it is essential to question people who watch cable television. Once the population is defined, the sample can be drawn from the members of that population. Prior to conducting the survey, the sample size (number of respondents) should be determined. The sample size should be large enough to provide a picture of the population and small enough to allow for efficient data gathering in a reasonable period of time. My experience has shown that for school children, 50 to 100 is a good range for the sample size.

One way to provide a sample that represents the population is to select the sample members in a random manner. If the questioning is to be done by telephone, then random telephone numbers can be generated from a random number table. If you do not have access to such a table, one can be obtained through any college that has a mathematics department. If the telephone exchanges in your area are 261, 266, 452, and 458, random four-digit numbers are drawn from the table and added to one of the exchanges to complete a telephone number. If the first four digits obtained from the table are 7215, then the first telephone number to be called is 261-7215. Flipping through the local phone directory and selecting numbers randomly, without looking at the names, can work as well.

A survey that is conducted by going door to door or by questioning people in front of a store or in a shopping mall is less likely to yield a random sample. Care should be taken not to restrict the sample to people known by the student. It should be emphasized at this point that random selection should be discussed with the class and that although randomness is vital in real-world research, the purposes of this project would be served even if the sample was biased.

The students are now ready to gather their data. They should be given a reasonable amount of time to conduct the survey. Once they have gathered their data, they should place the numbers in a frequency distribution chart (fig. 8.1) using two columns, one for measurement and one for frequency—the number of times the measurement was obtained. Notice that the sample size (98 in this instance) is also shown. The students should construct a bar graph to depict the results shown in this chart. Discuss with them the use of scale and color to emphasize results.

How do you rate Cheap Charlie's Discount House with respect to employee courtesy—excellent, good, fair, poor, or terrible?

Measurement	Frequency
5 (excellent)	10
4 (good)	11
3 (fair)	27
2 (poor)	31
1 (terrible)	19
	98

Fig. 8.1

Next they should complete their statistical computations. This could simply involve adding up the data and dividing by the sample size to obtain the mean. If the standard deviation is to be computed, a more formal representation should be used. This article will not consider standard deviation, since many teachers are not familiar with the statistic. Certainly the types of computation incorporated into the project will be a function of the teacher's knowledge and the ability level of the class.

When the students have completed a frequency distribution chart and a graph of the data, it is time to make some statistical computations. These computations serve two purposes. They of course provide a representation of the data, but they also demonstrate the use of arithmetic operations in a real-world situation. Wherever possible, computations should be done or verified with a calculator.

The *mean* (or average) is a single number that represents all the data. Averages, although they are not always means, are part of the experience of most students. Test grade averages, average height and weight by age, and averages in sports, such as average yards gained or average points scored per game, are some common examples.

The mean is obtained by adding all the measurements and dividing by the number of measurements. The number of measures, or sample size, is denoted by the symbol, n. Thus, in the survey on Cheap Charlie's, $n = 98$. Each measurement is multiplied by its frequency. For instance, in that survey ten people responded with a measurement of 5 (excellent), and so we multiply 10 times 5. The sum of these products is the sum of all the measurements. Dividing by n yields the mean. If the answer obtained for the mean requires rounding off, I usually have students round it to the tenths place. Computing the mean requires multiplication, addition, division, and rounding.

We are also interested in the proportion of the sample that falls in each category. The sample proportions are estimates of the population proportions. Proportions are best understood if they are reported as percentages. To obtain the percent, we find the relative frequency, a fraction whose numerator is the frequency and whose denominator is the sample size. The relative frequencies are converted to percentages. Thus, obtaining the proportions involves using a fraction to express a ratio and converting a fraction to a percent.

The results of the computations, again using the numbers in figure 8.1, are reported in table 8.1. All mathematical work should be checked on a computer if one is available that has an appropriate statistical program in its library.

The most important section of the survey report is the last part. Here the

TABLE 8.1
Descriptive Statistics

Measurement	Frequency	Frequency × Measurement	Relative Frequency	Proportion
5	10	50	10/98	10.2%
4	11	44	11/98	11.2%
3	27	81	27/98	27.6%
2	31	62	31/98	31.6%
1	19	19	19/98	19.4%
	98	256		100.0%

NOTE: Mean= 2.6

student must analyze, in writing, what the results of the study are. Comments should be made detailing the significance of the mean. Certainly a discussion of the estimated proportions should be included. Consequences of the results must be discussed. For example, if a report on a discount store's prices indicates public perception that the price level is about the same as in other stores, the student should bring forward suggestions for improving consumer opinion of the store's prices.

In our attempt to make what is learned in school relevant to life, one method is usually overlooked. By exposing the students' work to an interested audience, we cause that work to become relevant. A paper or a report that is graded by a teacher, admired by a parent, or even displayed on the bulletin board is not truly exposed to a real-world audience. The reward of having a poem or story published in a children's magazine is more meaningful to a student than receiving a grade of A +. The statistical survey can be a vehicle for making what is learned in school more relevant. Project reports that are considered well done are returned to the students for any corrections that might be needed. The corrected copy is mailed to a person or organization to whom the results would be interesting. A cover letter is included with the report, giving the age of the student, explaining the purpose of the project, and requesting a response. A survey on consumer attitude toward our local K-Mart was sent to the K-Mart regional vice-president for marketing. A report on smoking was sent to the state lung association. A survey on foreign policy was mailed to the district congressman. A report on telephones was sent to the public affairs manager for the local Bell operating unit.

Most of the reports that I mail out receive responses. These responses are displayed on a bulletin board to reward the authors of the reports and to make all students aware of the importance of the surveys. A letter from a local television station indicating that the student sampling of viewer attitude on local news programming replicated the professional analysis was received with much enthusiasm by the students.

Those students whose work on the reports is inferior must be made aware of the consequences of such work in the outside world. I tell these students that if they were being paid for this work (and there are many people who earn a comfortable living conducting marketing and political analysis), the poor quality of the work would require termination of employment. Also, any report receiving a poor grade is sent to the student's parents with a cover letter. The letter explains the purpose of the project, identifies the errors and omissions by the student, and asks for a response from the parent.

The statistical survey is an effective way to focus student activity on problem solving, to demonstrate the use of basic mathematical skills in relevant situations, and to incorporate calculators and computers. The communication of what has been done by the students through letters to businesses and government representatives builds an awareness in the community of what is being accomplished in the schools. Heightened awareness should increase public support for mathematical instruction. Students, parents, and taxpayers will not have to ask, "Why do we need to teach mathematics in school?" The need for mathematical skills and the integration of mathematics with other areas of learning will be demonstrated in the statistical survey, with relevance and importance apparent to participants and observers.

9

Estimation and Reasonableness of Results

Bob Underhill

ALTHOUGH everyone agrees that estimation is an important skill, many children and, indeed, many adults have poorly developed estimation skills. This article explores the concept of estimation and suggests some techniques for using estimation in addition, subtraction, multiplication, and division. Consider the following settings in which estimation skills are used:

"How much soda pop will we need for our group?"
"About 200 or 250 cans."

"How long is the room?"
"It's between 20 and 30 meters."

Such responses are commonplace. They help two people communicate by placing upper and lower bounds on the amount in question. They are intrinsically different from the following responses:

"How much soda pop will we need for our group?"
"About 225 cans."

"How long is the room?"
"It's about 25 meters."

In the latter statements, there must be some agreement between the conversationalists about the precision they will use; otherwise, they may not communicate fully.

The Importance of Context

When there are no rigid rules, how do two people communicate inexact numerical information? It is through a set of similar experiences in everyday living. Two farmers raise chickens and one says, "How many eggs do your hens lay?" The other responds, "About 35 dozen." These two farmers understand each other because they have talked with farmers and family members about eggs. Each is assuming "35 dozen each week" or "35 dozen each day." They do not need to say "each week" or "each day" if they have communicated like this in other situations. Local people know whether "each week" or "each day" is implied because of many such *contextual* conversations. If these are chicken farmers whose eggs are a major source of income, they probably mean "each day." But if they raise chickens for home use and sell their surplus eggs to friends and neighbors, they probably mean "each week."

Similarly, it is from placing such conversations in context that the farmers know whether "about 35 dozen" means "30 to 40 dozen" or "34 to 36 dozen." A stranger who drops in on the conversation will not know the meaning of "about 35 dozen," either in daily volume of egg production or in the range of egg production. For the stranger, the farmer must say something like "30 to 40 dozen each week." But the newcomer would not immediately know the local assumptions behind such a statement.

In scientific work, the need to be precise is probably greater than in everyday conversation, and so scientists have developed rules or procedures for interpreting scientific numerical information. The scientist uses significant digits and a rule of ± .5 of the last digit. Here are a few examples:

1.3 is assumed to be between 1.25 and 1.35

270 is assumed to be between 269.5 and 270.5

42.67 is assumed to be between 42.665 and 42.675

3600 is assumed to be between 3595 and 3605

Communicating Estimates

The need to communicate estimates is important because estimates often communicate a range or an average. The two farmers may have understood that 35 dozen was an *average* production figure or that 30 to 40 dozen was the *range* of egg production during the peak laying season. In either case, these figures are often much more indicative of the overall production than the figure for any single day.

Suppose the farmer asked instead, "What was your egg production yesterday?" If it was a poor day, an answer of "32 dozen" might be a bit misleading. If it was a particularly good day, an answer of "38 dozen" would be equally misleading. Therefore, the average egg production is a better figure because it takes into account some of the day-to-day variation in production.

The average is a more sophisticated concept than the range because it assumes a sum of production figures and a division by the number of days. To intuit a mean or average production figure, one needs to reflect on the production figures of several days and to notice the frequency of highs, lows, and middles. Study the following list: 37, 42, 29, 33, 40, 28, 35, 31. The probable steps in determining an answer might be these: (1) The high is 42, (2) the low is 28, (3) there are four lows, (4) there are two highs, (5) the middle is about 35, (6) the mean is below 35, (7) about 34. If these figures represent egg production, the farmer might say that he gets "about 34 dozen." It would be much easier to say that he gets "between 29 and 42 dozen." To say that he gets "about 34 dozen" doesn't adequately communicate the expected production to one who is unfamiliar with the data. To say "between 29 and 42" is much better for the novice.

Novices have a limited set of data from which to judge the impact of *means* or *averages*. The interpretation of an average depends greatly on the spread of the data. Study the three sets of data in figure 9.1.

	100		50		40
	100		50		40
	100		50		40
	0		50		40
	0		49		40
	0		51		100
Mean =	50	Mean =	50	Mean =	50
Range =	100	Range =	2	Range =	60

Fig. 9.1

In each set the mean is 50, but the ranges are 100, 2, and 60. "About 50" might be very misleading to novices who have had no experience with the data. With a value of 40, they might think, "That is close enough to 50." In the first and third sets, they would be correct, but in the second set they would be in serious error. Thus, novices would be better off to begin with the *range* to establish meaningful expectations or *reasonableness of answers*.

Teaching Estimation through Ranges

Certainly our students are novices with respect to the new content we teach. It seems reasonable, therefore, to begin work with *range* estimation. We already do this with many of the measurement skills we teach.

Range estimation in measurement

When we introduce students to measurement skills, we usually begin with length, and we approach length measurement concepts through readiness experiences. For example, we can use one or two pencils to determine the length of a desk (fig. 9.2). We then say that the desk is "between 8 and 9 pencils" long. Similarly, when we measure a room with a meterstick, we say that the length of the room is "between 11 and 12 meters" long (fig. 9.3).

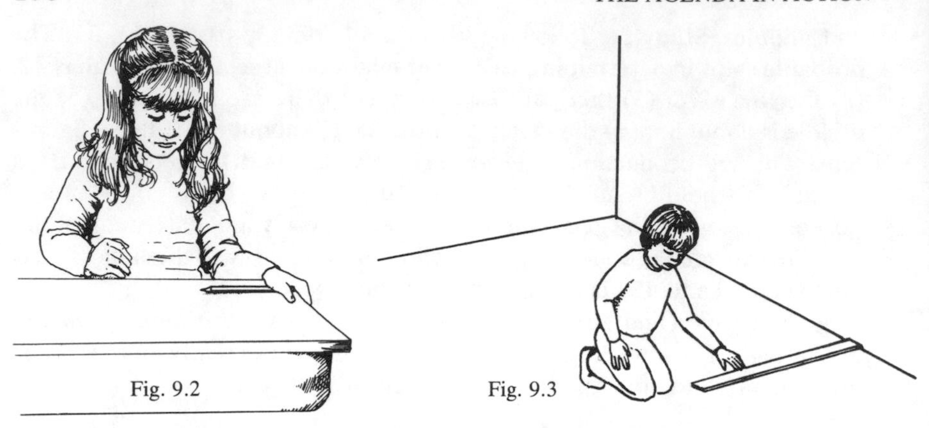

Fig. 9.2

Fig. 9.3

When we introduce area, we can follow a sequence such as that presented in figure 9.4.

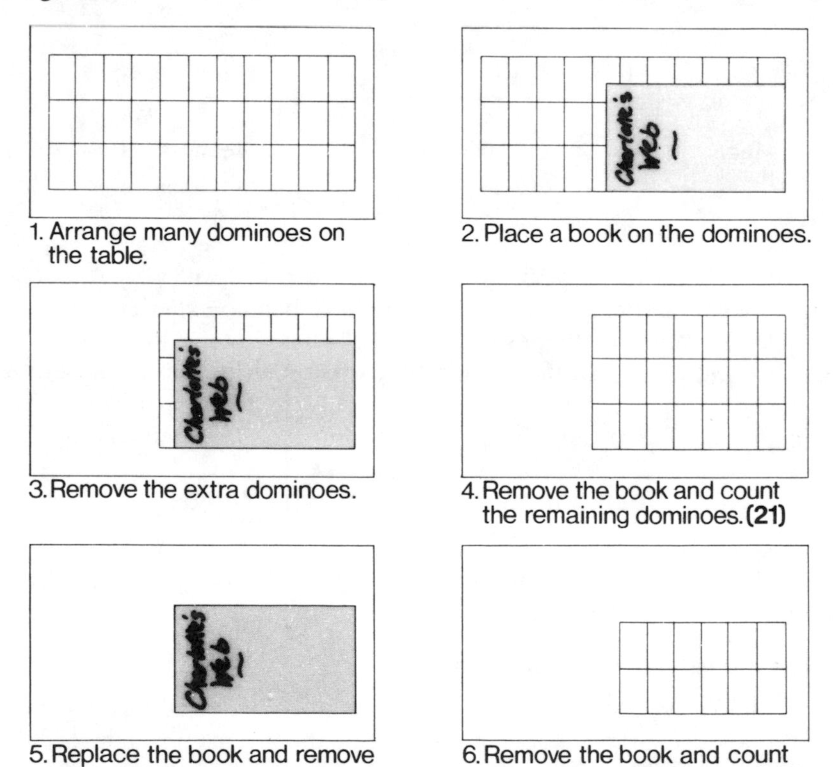

Fig. 9.4

In teaching weight, we can use a balance scale and add weights until the scale tips. If we have been adding 100-gram weights, for instance, we know between which two 100-gram multiples an object's actual weight falls. In figure 9.5, the can on the scale weighs between 200 and 300 grams.

Fig. 9.5

Range estimation in addition and multiplication

If we are teaching range estimation in measurement skills, it should be fairly easy to teach it in addition and multiplication, too. Study the addition examples in figure 9.6

Addition		
	26 + 38 =	
Round down		Round up
20 ←	**26** →	30
+30 ←	**+38** →	+40
50 <		< 70

Addition		
	$3\frac{5}{9} + 8\frac{3}{7} =$	
Round down		Round up
3 ←	$\mathbf{3\frac{5}{9}}$ →	4
+8 ←	$\mathbf{+8\frac{3}{7}}$ →	+9
11 <		< 13

Addition		
	0.062 + 0.134 =	
Round down		Round up
0.0 ←	**0.062** →	0.1
+0.1 ←	**+0.134** →	+0.2
0.1 <		< 0.3

Fig. 9.6

The process is quite straightforward. Simply round all addends *down* and add. Then round all addends *up* and add. The desired sum is between these two sums. The rounded-down sum and the rounded-up sum constitute the range within which the desired sum will lie.

Since multiplication can be interpreted as repeated addition, the range estimates are made similarly. The examples in figure 9.7 illustrate the similarity.

Multiplication		
$26 \times 38 =$		
Round down		Round up
20 ←	**26** →	30
×30 ←	**×38** →	×40
600 <	**988** <	1200

Multiplication		
$3\frac{5}{9} \times 8\frac{3}{7} =$		
Round down		Round up
3 ←	**$3\frac{5}{9}$** →	4
×8 ←	**×$8\frac{3}{7}$** →	×9
24 <	**$29\frac{61}{63}$** <	36

Multiplication		
$0.062 \times 0.134 =$		
Round down		Round up
0.0 ←	**0.062** →	0.1
×0.1 ←	**×0.134** →	×0.2
0 <	**0.008308** <	0.02

Fig. 9.7

As with addition, simply round both factors down and multiply; then round both factors up and multiply. The actual product will lie between these two values.

Range estimation in subtraction and division

Subtraction and division are inverse operations of addition and multiplication. We are working with one addend and a sum in subtraction and with one factor and a product in division. We cannot use the same approach we use with addition and multiplication; it will not work consistently. Study the examples in figure 9.8.

Subtraction		
Round down		Round up
500 ←	**562** →	600
−100 ←	**−145** →	−200
400	**417**	400
There is no range!		

Division		
Round down		Round up
200	←**204**→	300
10	←**17**→	20
20	**12**	15
12 is *not* between 15 and 20.		

Fig. 9.8

It can be demonstrated mathematically that the range method of estimation works with these inverse operations by rounding one number *up* and the other *down,* then the reverse.*

---

*If $a + b = c$, $b' > b$, and $c' < c$, then $a > c' - b'$.
If $a + b = c$, $b' < b$, and $c' > c$, then $a < c' - b'$.
If $a \times b = c$, $b' < b$, and $c' > c$, then $a < c'/b'$.
If $a \times b = c$, $b' > b$, and $c' < c$, then $a > c'/b'$.

Subtraction. To subtract, round one number (either one) up and the other number down, then subtract. Then reverse the rounding and subtract again. The order of rounding doesn't matter as long as you do *not* write in the < or > signs until you are finished with the estimation. Study the example in figure 9.9. It is done both ways.

Subtraction

70←	up ←	**63** →	down→	60
−10←	down←	**−14** →	up →	−20
60>			>	40

Subtraction

60←	down←	**63** →	up →	70
−20←	up ←	**−14** →	down→	−10
40<			<	60

Fig. 9.9

It is probably better to set no hard-and-fast rule for order, since it is quite simple to use < or > as appropriate. Illustrations for decimals and fractions appear in figure 9.10.

0.1←	down←	**0.168** →	up →	0.2
−0.1←	up ←	**−0.019** →	down→	−0.0
0.0<			<	0.2

10←	up ←	$\mathbf{9\frac{1}{6}}$ →	down →	9
− 4←	down ←	$\mathbf{-4\frac{4}{5}}$ →	up →	−5
6>			>	4

Fig. 9.10

Division. In division, the use of place-value names is most helpful. It is also quite helpful to use a basic facts range of difficulty. Study the examples in figure 9.11.

Division

$32\overline{)1753}$

17 hundreds	←down←	**1753** →	up →	18 hundreds
4 tens	← up ←	**32** →	down→	3 tens
4 tens				6 tens
40	<	**54+**	<	60

$\frac{100}{6}$	← up ←	$\frac{92.3}{6.2}$	$\mathbf{\frac{9.23}{0.62}}$	$\frac{92.3}{6.2}$	→ down →	$\frac{90}{7}$
					(up)	
$16\frac{2}{3}$	>		?		>	$12\frac{6}{7}$

(Denominators: 6 ← down ← 6.2; 6.2 → up → 7)

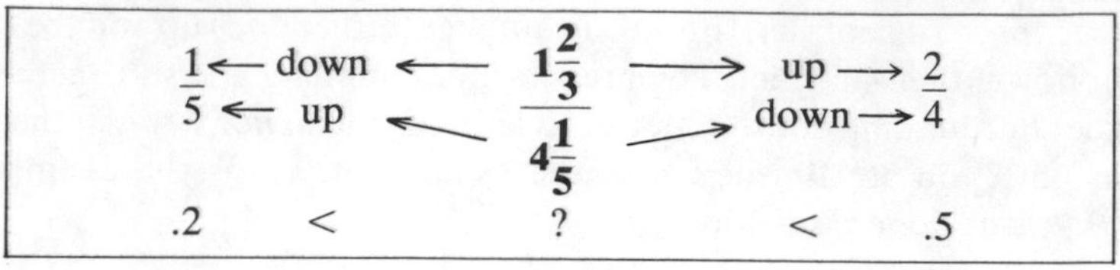

$$\frac{1}{5}\begin{matrix}\leftarrow \text{down} \\ \leftarrow \text{up}\end{matrix} \quad \begin{matrix}\longleftarrow \\ \longleftarrow\end{matrix} \quad \frac{1\frac{2}{3}}{4\frac{1}{5}} \quad \begin{matrix}\longrightarrow \\ \longrightarrow\end{matrix} \quad \begin{matrix}\text{up} \rightarrow \\ \text{down} \rightarrow\end{matrix}\frac{2}{4}$$

$$.2 \quad < \quad ? \quad < \quad .5$$

Fig. 9.11

Division, like subtraction, is a secondary operation, and so it is necessary to round up/down and down/up.

Summary

It is nearly always a logical and natural outcome that the measurements we make are bounded on either side by values whose accuracy varies according to the precision of the tool we use. The estimation strategies suggested here for addition, subtraction, multiplication, and division of whole and rational numbers use this range concept to help students understand or intuit the reasonableness of their answers by helping them establish similar upper and lower bounds.

10

Using the Front-Page News to Teach Mathematics

Betty Harmsen

THE search for up-to-date applications of mathematics principles, geared to the student's ability level, is often a difficult one. Textbooks become outdated so that students cannot relate to the problems they contain. This is especially true when older students doing remedial work are confronted with problems designed for younger students. A high school student cannot feel good about being asked to solve a problem such as this: "John's seventh-grade class has 27 students, and 3 of them are absent. What percent are present?" The attempt to create more appropriate problems is often frustrating, especially at the end of a day's duties when time and creativity are both in short supply. One solution to this problem comes from the front page of the newspaper, where there is a never-ending variety of "starters" for up-to-date and relevant problems.

Using problems based on news articles gives the satisfaction of knowing you are contributing to interdisciplinary efforts. (Even if the students don't figure out how to work the problem, they will be more informed on current events!) Newspaper problems are not only relevant for the student, they are fun for the teacher. You can discover whether they are paying attention with one like this:

> The Burlington Railroad owns one square mile of land on top of Mount St. Helens. Until recently they were paying $14.20 a year for fire protection. How much was that per acre?

Or capture their imagination with one like this:

> An Omaha biology teacher gave mouth-to-mouth resuscitation to 4 of his 42 jumping frogs who were left out in the cold and had to be thawed out. What percent of his frogs required mouth-to-mouth resuscitation?

News articles frequently contain large numbers, which will give much-needed practice in an area where most students are weak. It is advisable for each student to have a calculator to use in order to minimize frustration and keep them from getting sidetracked from the main task of deciding *what* to do in the problem. However, it is a valuable learning experience for students to discover that even with the calculator, they must learn some manual techniques to deal with numbers that won't fit into the calculator, such as $655 billion.

To write a problem, most of the information can be taken directly from the article, but occasionally some numbers need to be added to turn the information into a question. For example, if an article says that silver sold for $40.75 an ounce, you could ask how much two pounds of silver costs. Most problems should probably be kept fairly short, but you can develop skills in sorting out extraneous numbers with a few like this: "The ABC Mortgage Finance Fund is selling bonds at a lower rate of interest so that banks can make home loans at 10.75% interest rather than at the going rate of 14.25%. Suppose a person bought a $50 000 house and paid $10 000 down, taking a thirty-year mortgage for the remaining $40 000. The buyer would thus pay $375.39 each month on a 10.75% loan rather than $489.82 monthly if the loan were at 14.25%."

Many times one article can be the basis for several problems, or you can vary the question to fit the student's level.

Sample: A sterling silver coin has been issued to honor Queen Elizabeth. It is 38.6 millimeters in diameter and 28.28 grams in weight. It sells for $73 in U.S. currency and $86 in Canadian currency. Only 25 000 of these coins have been issued.

Possible questions:

- How much is the total cost of all 25 000 pieces in U.S. money?
- How many grams per millimeter is the coin?
- How many square millimeters in area?

- How much is a Canadian dollar worth in U.S. money?
- A U.S. dollar is worth what percent of a Canadian dollar?

Problems can be designed to deal with one concept at a time or with several concepts simultaneously.

Sample: A 2.5-inch-long gypsy moth caterpillar eats twelve square inches of foliage a day.

A single-concept question could be—

- How many square inches of foliage does the gypsy moth caterpillar eat in three days?

A multiple-concept question might be—

- If an orchard has 60 trees averaging 3000 square feet of foliage each, how many gypsy moth caterpillars would it take to completely consume the foliage in three days?

For long-term retention of the skills involved in solving these kinds of problems, I recommend that this type of work be given throughout the year, say one day every two weeks, rather than as a single two- or three-week unit. Because students need repeated practice in learning to discriminate among the different types of problems, it is advisable to mix several types of problems from the beginning and to give at least ten to fifteen problems for each assignment. I do not try for mastery the first time, since it is through repeated exposure to the large numbers that students learn to feel comfortable with them. Also, one should be very cautious about suggesting short cuts to students who aren't ready for them. Mathematics is not all intellectual knowledge, and some students need the physical experience of going through the steps several times before they really understand the problem.

One of the most powerful tools in elementary mathematics is the use of proportions. Newspapers are a natural for this, and you will have no trouble finding an ample quantity of "per" problems, measurement conversions, comparisons, and percents.

Sample: In Moscow, a new Fiat 124 costs 9200 rubles, which is about $13 000.

A possible question might be—

"How much is a dollar worth in rubles?"

Some students can do this type of problem intuitively, whereas others need to set up a proportion to know whether to multiply or divide and to avoid dividing upside down. If students have developed a habit of checking for reasonableness and if estimation is an integral part of the curriculum, they will more readily recognize any errors they do make. In setting up their proportion, some students will just naturally put rubles over rubles, and others will put rubles over dollars. If you show the class only *your* favorite way, without mentioning that there are several correct ways to write a proportion, you may confuse them and even destroy the understanding they thought they had. Individualize the method you give students whenever you can; they don't all need to do the problem the same way. They will feel better about their own mathematical ability if you build on their previous skills and their natural intuition and if you spend more time asking than telling.

The format I have found most helpful in setting up the percent proportion is as follows:

$$\frac{\text{Result}}{\text{Original number}} = \frac{\text{Percent}}{100}$$

Here I am using the word *result* in the same sense as *percentage*. Through your repeated questioning, students develop an intuitive feeling for what kind of quantity represents the original number and what kind represents the result. Ask such questions as "Which has to happen first, the sales or the sales tax?" and "Is the money in the bank the result of having interest or is the interest the result of having money in the bank?" Be sure to include problems where the result is larger than the original number; otherwise, students may come to believe that the original number is always just the larger of the two numbers.

It needs to be stressed repeatedly that the result number must "tie to" the percent. In other words, if the percent is the percent of increase, then the result must be the amount of increase; if the percent is the percent of commission, then the result must be the amount of commission, and so on.

Sample: The number of people who are between the ages of 15 and 24 is projected to decrease from 40 million in 1980 to 32 million by 1987.

Students will often put the number 32 million as the result number in the proportion, but if you have carefully taught them that the result must "tie to" the percent, they will understand when you say, "Let's see now, we are looking for the percent of decrease; is 32 million the amount of decrease?" It also helps to have the students actually write out the words in the proportion, as follows:

$$\frac{\text{8 million decrease}}{\text{40 million population in 1980}} = \frac{N\text{ percent decrease}}{\text{100\% population in 1980}}$$

You will want to give your students several types of percent problems in each assignment, but because some types occur more frequently than others in news articles, you may have to formulate your questions with care to get the variety you seek.

Sample: The price of gold was $461.50 an ounce on 7 June 1981 compared to an average of $614 in 1980.

The most obvious question to ask is "What percent decrease is this?" However, if you go ahead and figure the percent of decrease yourself (it's about 25%), then you can choose either of the following problems:

> The price of gold has decreased about 25% since 1980 and is now around $462 an ounce. Approximately what was the price in 1980?
>
> The price of gold has dropped about 25% from its 1980 average of $614. Approximately what is the price of gold now?

You could also subtract the dollar amounts to get the amount of decrease (about $152) and pose a question such as this: "The price of gold has dropped about 25%, or $152. What is the approximate price of gold now?" The difference between such questions may seem trivial to us, but it is not trivial to the student. Having problems of all types mixed together instead of a single repeated type can mean the difference between rote learning and real understanding.

I believe that if students develop these skills, their confidence in their mathematics ability will increase significantly. Picture Mary, one of your students, ten years from now. Mary is reading the newspaper and comes across an article that says, "Census figures released recently show that the population under 25 years of age has dropped to 117.8 million, or 42% of the total." Now suppose Mary has an idea in her head of what the population of the United States is, but she isn't sure how current that figure is, and maybe she has even been wondering lately how much it has increased. Here is the answer, so close and yet hidden from view. Will she still be wondering when she puts down the paper, or will she be able to jot down a few numbers and walk away with a sense of quiet satisfaction?

11

A Statistics Course to Lighten the Information Overload

Jim Swift

NORTH AMERICA has changed from an industrial society to an information society. In 1950, 65 percent of American workers had industrial jobs and 17 percent had information-related jobs in such areas as computing, teaching, public relations, television, books and magazines, marketing, advertising, and accounting. During the last thirty years this ratio has changed dramatically. Now over 50 percent of all workers are in information-related jobs and only 30 percent in industrial jobs. Our society has also grown more and more complex. One computer terminal is available for every forty-eight workers, and by 1986 we can expect one terminal for every ten workers. It is easy to see how such tools help with the prolific generation of information.

Consider the following analogy. Look at a stack of books and let it represent all information recorded in human history up to 1950. That stack was doubled in the two-year period of 1951 and 1952. By 1955, the duplication time had decreased to one year, and by 1980 it had shrunk to two months. Glover Anderson, chairman of the Canadian Telecommunications Carriers Association, described one of the symptoms of this information overload at a seminar in Vancouver, British Columbia. He asked his audience:

> How many of you have stared at a 500-page computer printout—one full of vital statistics—that is beyond your ability to understand, longer than time permits to read and more than your patience can endure?

It seems that we must now recognize a new basic skill, that of being able to make sense of numerical information. This is a skill demanded of anyone who reads newspapers or magazines. They often contain data, graphs, and surveys, and readers are continually being confronted with arguments that require the understanding of such numerical information.

The task of developing a curriculum for this information society has many components. One such component is to rid ourselves of the notion that statistics at the school level consists of either bar graphs and a few calculations or else an elective course for a few senior students. It is not so! The subjects of probability and statistics are a rich source of skills with which we can reach outside our classrooms, take data on a piece of the world, examine those data, and from them extract all the information we can. To pretend that such a process is dull, dry, and uninteresting is a lopsided look at learning. Fortunately, more and more teachers are recognizing the joys of statistics, and more and more students are discovering the vitality of statistics.

The *Agenda for Action* recognizes this trend toward an information society and the need to prepare children for it:

> There should be increased emphasis on such activities as—
>
> —locating and processing quantitative information;
>
> —collecting data;
>
> —organizing and presenting data;
>
> —interpreting data;
>
> —drawing inferences and predicting from data. (p. 7)
>
> [Students must be equipped with] mathematical methods that support the full range of problem solving, including . . . methods of gathering, organizing, and interpreting information, drawing and testing inferences from data, and communicating results. (p. 3)
>
> The mathematics curriculum should [include] presentation through activities, graphic models, . . . schematic diagrams, simulation of realistic situations, and interaction with computer programs. (p. 4)
>
> For those whose formal education will end with high school, the needs of citizen and consumer for increasing mathematical sophistication dictate a collection of courses based on consumer and career needs, computer literacy, and quantitative literacy. (p. 18)
>
> Consumer mathematics should develop a broader quantitative literacy and should consist primarily of work in informal statistics, such as organizing and interpreting quantitative information. (p. 21)

It is to the implementation of recommendations such as these that this article is addressed.

Statistics, a Basic Skill in an Information Society

A reliance on numerical information is common to all forms of decision making. A television commercial quoting a marketing survey on two brands of razors stated that 90 percent of the people who expressed a choice thought that brand A was better than, or no different from, brand X. This was thought by many to be a powerful argument for the purchase of brand A. But we were not told the breakdown of the 90 percent. What percentage of the people questioned could say the same thing about brand X—that it was better than, or no different from, brand A? This question was asked in British Columbia's recent mathematics assessment of grade 12 students, and only 14 percent answered it correctly. Even when we allow for the difficult wording of the question, this is a poor showing. But the words are taken from an actual commercial. Many mathematics teachers, even, agree with the answer 10 percent without giving thought to the overlap. All we know from the information given is that 10 percent of those questioned prefer brand X. We do not know how many thought there was no difference. It might be as high as 90 percent. The percentage who thought brand X was better than or no different from brand A could be as much as 100 percent!

A cursory glance through a newspaper will reveal many uses of the word *median:* "The median house price in Vancouver, B.C., is $150 000." "The median salary of the teachers is $29 241." Yet on the assessment mentioned earlier, only 30 percent of grade 12 students in British Columbia could find the median of 2, 2, 2, 2, 3, 4, 5, 7, and 8.

The day the Consumer Price Index figures are released is the busiest day of the month at the Statistics Canada User Services offices across the country. It is clear from the inquiries that many people do not understand the index, but it is also clear that the CPI is important to a large part of the population. But look at the partial explanation of the CPI in figure 11.1.

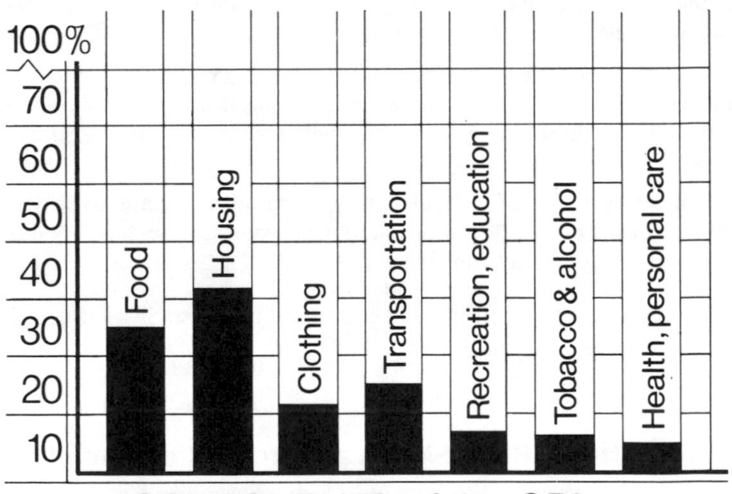

If a family's food costs or housing costs increase by 10 percent it does not mean the family salary must also go up by 10 percent. For instance this graph shows that the average family budget includes 27 percent spent on food which means that the 11 percent food cost increase over the past year is really 11 percent of 27 percent or approximately three percent of the overall budget.

Fig. 11.1. Reprinted with permission from the *Nanaimo Free Press,* October 1977

How many adults, let alone your students, would follow that argument? There are very few places in the curriculum where the idea of an index is taught. This prompted Gail Lenoski, director of the Statistics Canada user services division in Vancouver, to develop the unit described in "The Student Price Index," an article in NCTM's 1981 Yearbook, *Teaching Statistics and Probability*.

Surveys and opinion polls are also frequently misinterpreted. Students should be made aware, for example, that any opinion poll is but an estimate of the population's opinion. Consequently, rises and falls indicated by a poll should be ignored unless they are much more than the sampling error.

Papers like the *New York Times* are a fine source of clippings. Each year my statistics class spends the first few days collecting clippings that illustrate the wide use of statistical and probabilistic arguments. We continue this practice throughout the course, since it illuminates the course content with applications from everyday reading.

Students soon realize the depth of knowledge required to make sense of items like *average* or *mean* and (increasingly common) *median;* the laws of probability; the language of chance and odds; reading trends from time series graphs, and informal extrapolation from such graphs; probabilities associated with predictions, health risks, contraceptive failure, and so on; understanding that a sample can be used to give information about a population; how a sample is collected; and the uses of the census and other government surveys.

The list can go on interminably. Such topics must not be reserved for the grade 12 elective course, important though that is. They should become the basis of the consumer mathematics courses and be given a firm place in the mainstream of the mathematics curriculum.

I firmly believe that teachers should start by compiling their own collection of newspaper cuttings and asking questions about the information they contain. Alternatively, get the students to ask their own questions about the information. The Grand Valley Mathematics Association of Waterloo, Ontario, has published such a collection of clippings and questions under the title *Lies, Damn Lies and Statistics.* It was compiled by Steve Brown and Jock MacKay of the University of Waterloo and is a fine example of the possibilities of this approach.

Once the collection of clippings is under way, it soon becomes clear what relevant topics must be incorporated into the curriculum and at what depth the topics must be covered. Soon students will be exploring their own problems and realize that problem solving is not always confined to algebraic word problems.

Problem Solving: More than Solving Algebraic Word Problems

The subject of problem solving is identified in the *Agenda for Action* as the intended focus of school mathematics in the 1980s. The questions traditionally associated with problem solving have often been restricted to the word problems associated with an algebra course. The *Agenda,* however, makes it clear that this is a narrow view of problem solving. The strategies listed on pages 2 and 3 of the *Agenda* go beyond using mathematical notation to describe real-world relationships and include the skills of data analysis. The skills and strategies associated with examining data must also be developed as problem-solving strategies.

To illustrate some of these strategies in action, let me describe some of the problems brought to a high school statistics class by students or their parents.

Problem: The rapid expansion of a retail business area in a small town

The town of Nanaimo, in British Columbia, is experiencing a phenomenal growth in the amount of its retail shopping space. Several people, including a newspaper editor and a parent who owned a small retail business, suggested that this was a topic worth exploring. So a student went to the planning department at City Hall for its report, "Profile Nanaimo." After some investigation the data in figure 11.2 were collected.

Area of retail shopping space in September 1979		882 000 sq. ft.
Area for which building permits have been granted (fall 1979)		2 201 000 sq. ft.
Area of retail shopping space in December 1980		2 152 000 sq. ft.
Total retail revenue in 1979		$187 000 000
Revenue per square foot of retail space		$196
	1979	**1981 (est.)**
Population in the primary trading area (including Nanaimo)	72 000	79 700
Population in the secondary trading area	54 900	59 100
Population in the tertiary trading area	60 800	67 000
Annual revenue required to maintain the additional retail space		$400 million
Annual revenue per family (3.2 people) needed in 1981 to maintain the additional space		
	primary area	$11 600
(a weighted average)	secondary area	$ 3 900
	tertiary area	$ 1 900
Annual retail expenditure per family in 1979		$12 233

Fig. 11.2

Since no expansion of the population on the scale needed to justify this increase seems to be planned for the next five years, these statistics raise far more questions than they answer. But this simply gives us more problems to solve—more data to collect. The most interesting question is, of course, what further information do the developers have in their possession indicating that their investment will succeed and not merely lead to a chain of bankruptcies?

The student doing this investigation was pleased to see his findings confirmed in a Canadian Press report a few weeks later (see fig. 11.3).

NANAIMO, B.C. (CP)—Like embarassed guests who arrive early and find their hostess still soaking in the bathtub, the national department stores have landed with a thud on Nanaimo's doorstep.

This Vancouver Island community of 44,400 is confidently looking forward to an economic boom that will bring free-spending workers flocking into shops and malls. The trouble is that the shopping malls arrived well before the boom.

Already the city has 16 square feet of retail space per person, four times the national average. And two more major shopping centers are coming.

Local market analysts calculate, on the basis that stores nationally need annual sales of $390 a square foot to break even, that a Nanaimo family of four will have to spend $19,200 a year in the stores for everyone to avoid losses. That spending would be over and above the family's expenditure on rent, fuel, insurance and similar costs.

FEELING PINCH

City statistics show more than 2.2 million square feet of existing, planned or proposed shopping-centre space. That total, which excludes independent stores represents more than 50 square feet for each city resident and more than 35 for the 61,900 in the primary trading area.

Advertising salesmen who have to pry money out of the older downtown merchants report that "local merchants don't have two cents to rub together."

At least one bank branch manager is tightening credit, guarding against a possible epidemic of small-business failures.

In an interview, Mayor Frank Ney said the rush of shopping-center developers was clearly "overkill."

City council debated whether to delay some projects but decided on a 5-to-4 vote to welcome everyone and let the free market sort things out.

Ney, who voted with the minority, said the smaller stores are his concern since "the big majors will survive."

The mayor, however, emphatically refuses to take an "I told you so" attitude. "Now that the decision has been made, I don't want to look at things negatively," he said. "I want to look at things positively."

FUTURE PROMISING

And, he continued, there is plenty of room in Nanaimo for positive thinking.

The city, indeed, has a broad base of existing and potential economic activity.

Fig. 11.3. Reprinted with permission from the Nanaimo *Daily Free Press,* 19 March 1980.

Problem: Planning a solar home

Another student knew someone who was planning a solar home in Nanaimo and wanted to find out how much supplementary heat would be needed. This is related to the number of heating degree days (HDDs) in the year. Imagine that the temperature remained at exactly sixty-seven degrees Fahrenheit for twenty-four hours. This is just one degree below the reference temperature of sixty-eight degrees Fahrenheit (20°C) and represents one HDD. So a mean temperature of thirty-two degrees Fahrenheit for ten days would represent $10 \times (68 - 32) = 360$ HDD.

A visit to the local airport produced monthly data for the previous ten years. The student noticed this strange set of figures pointing to a decline in the number of heating degree days during December over that period:

Year	1971	1972	1973	1974	1975	1976	1977	1978	1979	1980
HDD	983	951	763	765	473	412	500	517	413	414

To look for a cyclic effect, we obtained data since 1951 and then plotted it on a computer. The computer used a method of smoothing out the fluctuations month by month to reveal an overall trend (fig. 11.4).

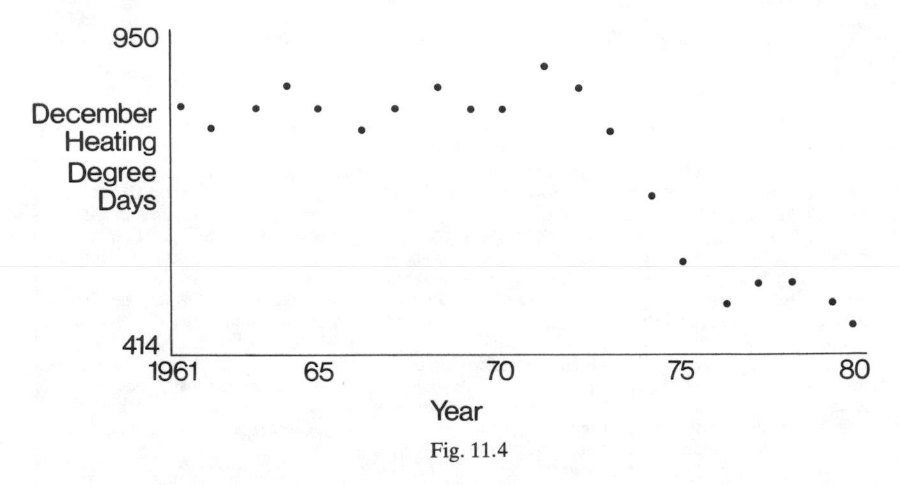

Fig. 11.4

The graph did not suggest a cyclic pattern in the decline, but it did emphasize a pronounced reduction in the number of heating degree days, from a mean of about 900 before 1975 to a mean of under 500 after 1975. Had there been a major change in the weather patterns in 1975? This time, given the clue of 900 and 500, the problem was easily solved. A phone call to the airport confirmed the solution that we suspected. We leave it for the reader to discover but will give it at the end of the article.

Problem: List and closing prices as related to the time taken to close a house sale

The student's mother, a realtor, wanted to see if there was any connection between the time taken to sell a house and the difference between the list and closing prices. The student thought that the longer it took to sell a house, the greater would be the difference between the list and selling prices. We explored the data she collected with a microcomputer, using the exploratory data analysis techniques described in *Interactive Data Analysis* (MacNeill 1980). Her opinions were not confirmed, as can be seen from the scatter diagram in figure 11.5.

Problem: Coaching a minor league hockey team

The student wanted to demonstrate to his hockey team some of the things

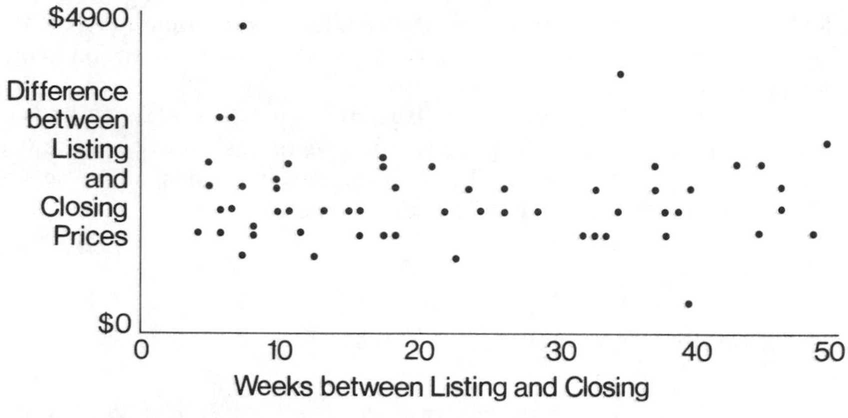

Fig. 11.5

that seemed to be connected with winning a game. He collected data from National Hockey League games indicating that the odds of winning were two to one in favor of the team that scored the first goal, and five to one in favor of a team that scored a short-handed goal.

In this course, students learned that certain problems can be solved only after suitable data have been collected. Such problems are commonplace in local, provincial, and federal government; in business; in such controversial matters as the effects of the nuclear accident at Three Mile Island and the health hazards of radiation, food additives, and pollution; and in the effects of climatic changes on food production. The list is endless. One of the keys to success in this kind of problem solving is student interest in the subject matter. Given this interest, students will develop the habit of looking at the world with the tools of mathematics at their disposal, an admirable objective in any curriculum.

Developing a Curriculum for the Information Age

The phrase "to seek out and find" from "Star Trek" contains the essence of our task in developing a curriculum for the information age. Where do we begin that task? We can begin with two objectives:

1. Encourage students to develop an area of interest and feed that interest with the collection of information.
2. Explore with them ways of examining the data, not as ends in themselves, but as tools to illuminate the data.

Neither of those objectives is new nor does either require a radically different view of teaching. Elementary school teachers have been using similar objectives for a long time. What might be considered different are some of the skills of exploring data. Although these have been covered more thoroughly in materials from the ASA/NCTM Joint Committee on the Curriculum in Statistics and Probability, it is worth drawing attention to one such skill here.

One of the keys to exploring data is to have quick and easy methods of looking at the entire collection. Quite often we are asked to compare two sets of data. Figure 11.6 shows the birth weights, in pounds, of two sets of girls born between 1961 and 1972 in San Francisco.

A

6.4	5.4	6.4	6.1	7.5	6.4	6.4	6.5	8.9	7.8	8.1	6.1	6.8
6.4	5.8	6.2	6.0	6.0	6.8	6.4	7.6	7.9	5.9	6.7	4.1	

B

7.9	9.5	7.1	7.6	6.4	6.2	7.4	7.1	5.3	5.4	6.8	7.0	8.1
6.3	8.2	9.7	8.3	7.8	7.2	7.4	7.8	5.6	6.8	5.8	7.2	

Fig. 11.6

No clear picture emerges from such a table. A better picture comes from a *stem-and-leaf plot.* Consider the first number in sample A, 6.4. We split the number into two parts, 6 (called the stem) and 4 (called the leaf). The stems in sample A are 4 through 8. We write these in a column. On the line for stem 6 we put the leaf 4. The second number is 5.4. The leaf 4 is put on the line for the stem 5. The third number is 6.4 again, and its leaf is put next to the first leaf (see fig. 11.7). When this is completed for sample A, each row is put into

4	
5	4
6	44
7	
8	

Fig. 11.7

ascending order. The complete diagrams for the two samples are shown in figure 11.8.

Group A

4	1
5	489
6	001124444445788
7	5689
8	19

Group B

5	3468
6	23488
7	01122446889
8	123
9	57

Fig. 11.8

Combining the two plots side by side (fig. 11.9) shows a much clearer pattern. Now it is obvious that the weights of group A are, on the average, less than those in group B.

Group A		Group B
1	4	
984	5	3468
887544444421100	6	23488
9865	7	01122446889
91	8	123
	9	57

Fig. 11.9

In such a situation, we should encourage students to ask such questions as "Were the two groups different in any way?" The answer in this case is yes. One of the differences is that the mothers in the group A sample were all smoking twenty or more cigarettes a day at the time the child was born. The mothers in the group B sample had never smoked. This example frequently prompts some students to gather further data from the study (Hodges, Krech, and Crutchfield 1975). For example, there are patterns to be examined relating to the birth weights for children of mothers over thirty-five and of mothers under twenty-five. A connecting link with the first example would be smoking habits of older mothers versus younger mothers.

A simple plot like this is not time-consuming and can be used in many different ways to illuminate data. The important goal is not so much that students learn to draw these kinds of plots but that by using them, they learn the *habit* of looking at data in an exploratory way.

Beginning this habit early in their school life is not something new to a large number of children. Many of them have already acquired the habit of counting things, from telephone poles on automobile trips to the number of steps they take to walk to school. In the curriculum of the 1980s, this is a habit to be recognized, encouraged, and developed. Not only will they develop a strong numerical interest, but the groundwork will be firmly laid for the new breed of specialist—the generalist who can extract from a set of data all its information and write summary reports for people who haven't the ability, time, or inclination to do so. The famous statistician R. A. Fisher refined this skill to a considerable degree and had insights that continually surprised his fellow workers. Here, then, is yet another challenge to mathematics teachers in the 1980s: Can we train ourselves and encourage our students to feel milking data for all it can provide can be an enjoyable and profitable skill?

Note: The apparent decline in heating degree days is the result of a change to the metric system—from Fahrenheit to Celsius.

REFERENCES

Hodges, J. F., D. Krech, and R. S. Crutchfield. *STATLAB—an Empirical Introduction to Statistics*. New York: McGraw-Hill, 1975.

MacNeill, Donald R. *Interactive Data Analysis*. New York: John Wiley & Sons, 1977.

12

Back to the Basics: One District's Response

Richard E. Cowan

In 1976, Roanoke Rapids City Schools in North Carolina began to develop a program and a process that would culminate in implementing Recommendation 2 from *An Agenda for Action.*

Everyone was talking about "back to the basics," and we wanted to be certain that the rallying cry was not detrimental to the total mathematics curriculum. We knew that basic skills should involve more than just computation. Therefore we undertook, with the help of Title IV-C funds, to determine which basic skills in mathematics are necessary for daily functioning in our community.

One of the lessons learned from the modern mathematics movement is that in order to effect curriculum change there must be involvement from a broad base of the population, including the lay public and teachers. With this thought in mind, we had a group of teachers choose eighty-six mathematics tasks that might be needed in normal functioning. Our staff designed several questionnaires, one sent to area employers asking which of these eighty-six tasks were needed for employment and another sent to the school patrons asking which of the eighty-six tasks they performed in daily functioning. On the basis of the replies, the staff chose fifty-four tasks and translated them into fifty-four skills that a high school graduate would need to function successfully as an adult. The staff then sent this list of fifty-four skills to over two hundred mathematics educators throughout the United States and Canada. These educators indicated the percent of their students who would be likely to need each skill, and a compilation of these percentages was used to order the skills from one to fifty-four. The process of selecting these skills is described in detail in the *Mathematics Teacher* (Cowan and Clary 1978). The skills are listed in ordered form in figure 12.1.

The next phase of the project was to determine minimal objectives for each of the skills. We decided to establish a continuous-progress K–8 curriculum. The skills to be taught at each grade level were then selected, and minimum objectives were written for these skills. The staff decided that

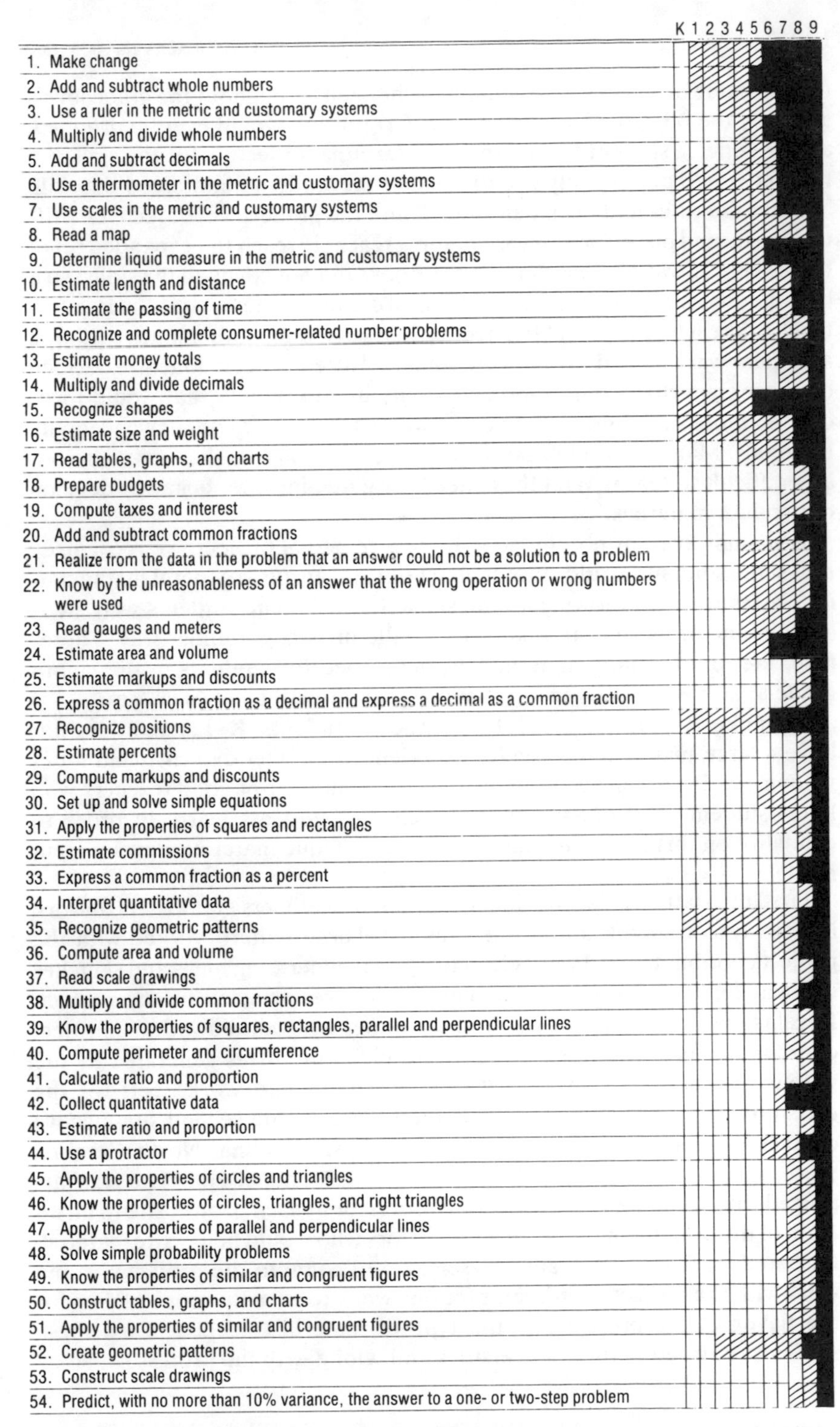

Fig. 12.1. Scope and sequence: The hatched area indicates the grade levels where the skill is developed; the black area indicates the grade levels where the skill is maintained.

(*a*) the name of a skill would remain unchanged throughout the grade levels and only the objectives would change, (*b*) no skill would ever be dropped, and (*c*) once a student had learned the terminal objective for the skill, the teacher would periodically test to see that the student maintained the skill.

After the skills had been determined and the grade-level objectives written, we turned our efforts toward helping teachers teach the expanded set of basic skills. It was obvious that the textbooks did not cover all the skills; so it became necessary to provide material for the teacher's use. Since every teacher at each grade level has access to at least one mathematics textbook and there was no need to duplicate what had already been done, we decided on a program that would supplement any basal text and that would focus only on those skills and objectives for which teachers did not have adequate materials. Material would also be provided for concepts that are traditionally difficult to teach. With these needs as guidelines, we began to develop curriculum materials.

Although this was not intended to be a remedial project, we recognized that some students would enter high school and enroll in general mathematics classes without knowing all the skills. In fact, some students were deficient in one or more of the skills before the project got under way. Therefore, the first teaching material was developed for students in the eighth grade and those already in high school general mathematics classes.

Roanoke Rapids is a small district (3000 students, K–12) and does not have a large mathematics faculty experienced in extensive curriculum development. It does have, as all districts do, a number of good teachers who know students and know what will help the students learn. In order to capitalize on this asset, our staff employed a unique materials development model.

First, the staff hired professional materials developers and writers to work in pairs for one week during the summer. These writers worked with the project coordinator and were given the job of generating ideas and sketches of ideas for teaching the mathematics objectives selected by the project. Local teachers were hired later in the summer and worked with the project coordinator for three weeks. They were given the ideas generated by the writers and asked to develop student materials and worksheets for the classroom. We did this for two summers. During the second summer we contacted three universities in the area—East Carolina, North Carolina State, and the University of North Carolina at Chapel Hill. The staff arranged with a mathematics education professor at each project site to hire a team of three mathematics education majors who would help write material. One student from each team was selected to come to the project site and work with the project coordinator for one week to get a thorough feeling for the flavor of the project and the type of materials needed. Then these students returned to their universities and, along with the professors, acted as team leaders for a week of developing and writing unique approaches for teaching the skills. All the material thus generated was again given to a team of local teachers at the appropriate grade levels, whose job was to forge the ideas into classroom materials for their students. In this way, the creative ideas of professional writers and fresh new ideas of university students were interjected into the project, and yet they were tempered with the experience and wisdom of local teachers who knew what would work with their students. The end result is teacher-created and teacher-written material that is being used by the teachers because they have developed it. We in the school system feel that this is a very cost-effective method of developing curriculum materials and that it should be used on a wider basis.

Early in the project, our staff decided that it was time to let teachers do what they are hired to do and, presumably, what they do best—teach. The project wanted to develop a teacher-centered curriculum that would provide help where help was needed but would not cause the teachers to sink in a quagmire of testing and paper work that would leave them no time to teach.

To address the goal of Recommendation 2, we needed to provide the teachers with two things of paramount importance: (*a*) good instructional materials and, perhaps even more important, (*b*) a method of assessing the new skills. Teachers want and deserve a method of assessing what they teach. A thorough search of the published instruments failed to turn up any that had a large enough intersection with our set of skills to be useful. It was therefore necessary to develop our own tests.

In keeping with the project's philosophy of a teacher-centered curriculum, the tests had to be useful to the teacher but not become the focus of the curriculum. The staff developed a diagnostic test for each grade level to test a sample of the skills taught. In addition, two mastery-level tests were developed for every skill at each grade level. These were designed to be used during the teaching of the skills—one to be administered at the introduction of the skill and one to determine the attainment level achieved at the completion of the instruction. All the tests were provided as optional help for the teacher. The teachers' judgment and their professional assessment of student learning were to be the final criteria used in evaluating students' accomplishments.

Continuous-progress charts for recording the attainment of the skills for individual students were developed for the skills at each grade level, with provisions for the teachers to indicate that the student is maintaining those skills for which the terminal objective has been achieved. Class-profile charts were developed to help teachers plan for the entire class.

In order to give teachers maximum help in meeting the students' individual needs, our staff developed the materials on a nongraded basis and packaged them by levels. Level 1 contains the material designed to be used with K–2 students. The material is organized by skill with all the K–2 material included in the skill package. No grade-level indication appears on the sheets, and so students do not know when they are working on grade level. All material can be used for individual help, small-group instruction, or with the whole class.

Level 2 material is to be used with grades 3–5, and it, too, is organized by skill and is to be reproduced for the students' use. Level 3 material is to be used with students in grades 6–8 and high school students who need remedial help.

The scope-and-sequence chart in figure 12.1 shows, with hatch marks, the grade levels where a skill is introduced and developed. After the skill has been achieved, it is maintained as indicated by the solid shading throughout the remainder of the sequence. Remember that the skill name does not change but that the grade-level objective is different for each grade.

Among NCTM's recommended actions for implementing Recommendation 2 are the following:

- Including among the basics all ten basic skill areas identified by the National Council of Supervisors of Mathematics in its "Position Paper on Basic Skills"
- Changing the priorities and emphases in the instructional program to reflect the expanded concept of basic skills
- Incorporating estimation activities into all areas of the program on a regular and sustaining basis
- Considering as basic the higher-order mental processes of logical reasoning, information processing, and decision making

The Roanoke Rapids project has attempted all these recommended actions. One-sixth of the skills are estimation skills with real-life applications, such as estimating money totals, estimating the passing of time, and estimating area and volume. The activities prescribed in the material depict real-life situations. For example, under estimating money totals, students are given problems like the following:

> **Mary bought a pair of socks for $2.43, a hat for $4.65, and a record for $6.35. If she has a five-dollar bill and a ten-dollar bill, will she be able to pay for her purchases?**
>
> **John bought groceries for $11.36. He gave the clerk a five-dollar bill, a ten-dollar bill, and two quarters. What is his change?**

Under estimating area and volume, the students look at objects in the room, select one as if they were going to pack it for mailing, and then choose a box in which the object would fit.

The problems under estimating the passing of time involve estimating elapsed time both within a time zone and across two or more time zones, such as those involving airline schedules, bus schedules, and so on.

Teachers are encouraged to interact constantly with the students concerning the problems at hand. Much of the student material is designed to be presented verbally in the form of a discussion or role playing. Since the material is supplemental, it seems to provide a relaxed atmosphere and open the door to student participation. In more than one instance, students have not recognized the project material as a part of the mathematics curriculum; they see it as a separate activity for fun and often comment, "Oh boy, no math today!" Students gradually become more willing to take a chance in making guesses and offering solutions to mathematical problems. Much of the material requires verbal responses by the students. The estimating often deals with *how far, how long,* or *how much.* Encouraged to respond verbally, the students are given the opportunity to defend their choices and explain their thinking. Incidentally, we found that the most effective way to force students to estimate was to place time limits that do not allow for computation or measurement. Given the opportunity, students seem to prefer the correct, exact answer to an estimate.

Students are given many opportunities to use logical reasoning and information processing in decision making. One of the better examples is contained in the level 3 material under the skill of reading tables, graphs, and charts. In this section the students are given the task of helping a family move from one location to another, basing their decisions on a considerable amount of information presented over the course of the unit in the form of tables, graphs, and charts. The unit takes about two weeks to teach and culminates with the family making a choice based on the data. Climate, distance, and opportunity for sports participation are the primary factors considered. Students seem to enjoy this unit, in which they must employ logical reasoning, information processing, and decision making, all while they are reading tables, graphs, and charts.

One of the most difficult areas to implement among those recommended by the National Council of Supervisors of Mathematics is reasonableness of answers. Everyone acknowledges that it is hard to teach children to examine their answers for reasonableness, but no one seems to have done anything about solving the problem. The approach that the Roanoke Rapids project takes is to separate the process of determining a reasonable answer from the process of determining the answer. Once students find an answer, they consider that they have finished. They are not going to do something "extra," such as determining whether the answer is reasonable or not, with the same intensity or interest as finding the answer. The materials include a wide variety of problems with answers already determined and give the student one task: deciding if the answers are reasonable. This has been extremely successful. A variety of presentations are used. Some are worksheets of computation problems that include answers, some are worksheets of "story problems" already worked out with the work shown, and some are sheets containing answers supposedly arrived at through estimation. In all of them, the students have only to determine if the answers are reasonable.

A second level of involvement is provided by giving the students word problems that are worked out and instructing them (*a*) to determine if the answer is reasonable, and (*b*) to find what was done incorrectly if it is not reasonable. This forces the student to check the work on a problem but still eliminates the double task of working it and then checking. All of this has contributed to the students' awareness of what constitutes a reasonable answer, an awareness that has transferred to other units. This skill is taught beginning in the fourth grade and continued throughout the project.

The project has been validated through the Education Department's Identification, Validation, Dissemination (IVD) process and is currently being disseminated to other school districts in North Carolina. Thirteen school districts in addition to Roanoke Rapids have used various portions of the project material to instruct approximately eight thousand students. For evaluation purposes the twenty classroom units using the material were matched with control groups. Eighteen of the twenty project units showed greater growth than their respective control groups. In only one instance did the control group show a significant gain over the experimental group, and in every instance the experimental groups showed positive gain over the course of the year. Two control groups actually regressed during the course of the year.

To implement Recommendation 2, teachers must be provided with materials to teach the additional skills and must also have a method to assess their students' accomplishment of these new skills. It is unfortunate, but too often true, that if an item appears on the test, it will be taught. Hence, for Recommendation 2 to be implemented on a widespread basis, test makers must be convinced to include test items that will assess skills other than those now being tested. However, Roanoke Rapids City Schools has been successful in implementing this recommendation and we would encourage others to join in improving both the teaching materials and the tests thus far developed.

REFERENCE

Cowan, Richard E., and Robert C. Clary. "Identifying and Teaching Essential Mathematical Skills—ITEMS." *Mathematics Teacher* 71 (February 1978): 130–33.

Recommendation 3

MATHEMATICS PROGRAMS MUST TAKE FULL ADVANTAGE OF THE POWER OF CALCULATORS AND COMPUTERS AT ALL GRADE LEVELS

13

Teaching Place Value with the Calculator

Ray Kurtz

TEACHERS start early in the first grade and continue in subsequent grades showing and telling students that our number system has the wonderful element of place value. Place value makes it possible to use only the digits 0–9 to represent numbers as large as the federal budget and even larger. This characteristic is not shared by such systems as those of the ancient Egyptians and Romans.

But a child's first encounter with place value can be critical. It is extremely difficult for some children to understand, for instance, that the 1 before the 0 in 10 means something far different from the 1 by itself.

Any inexpensive calculator can be a most helpful aid in teaching and reinforcing place-value concepts. The most elementary activity consists of pressing the [1], followed by [+], followed by [=], [=], [=], . . . , until a 9 appears in the ones place. The calculator then automatically begins using the tens column when the next 1 is added. Children will need to repeat this process a number of times, watching the ones place "fill up" and another "ten" appear in the tens place, until a 9 appears in the tens place. When the tens place is "filled up," a 1 will appear in the hundreds column. This shows as clearly as possible the trading of 10 tens for 1 hundred.

Whole Numbers

Even in kindergarten, children work with numerals up to 10 or larger. First-grade children work with numerals through 200, and second-grade children work with four-place numerals. Third graders encounter six-digit numerals, and fourth graders read and write seven-digit numerals. It is assumed that teachers of these grades will continue to use such time-proved aids for teaching place value as the abacus, place-value charts, bundles of sticks, and the hundred board. The calculator can be a valuable addition to these standbys in all the grades.

Let's first look at the first-grade objective of identifying ones, tens, and hundreds in a numeral such as 176. The teacher should prepare a set of cards (fig. 13.1). When the cards are put together, the numeral 176 appears (fig. 13.2). The teacher can put the cards together and take them apart a few times to help the children develop the concept of ones, tens, and hundreds.

Fig. 13.1

Fig. 13.2

After students have developed some understanding of place value, the calculator can be used to reinforce these concepts. The following activities have proved excellent in helping impart an understanding of place value.

"Whole number teardown"

In the game "whole number teardown," player A writes any three-digit numeral (four or more digits for older students) on a paper, enters it in the calculator, and then hands the calculator to player B. Player A tells player B to take away *all* the tens (or hundreds or ones). B enters [−] and the number necessary to follow A's directions and then pushes [=]. If this is done correctly, B scores one point. (The equal key may be pushed only one time each round.) In the next round, B records a number, enters it, and gives A a command to take away all the ones, tens, or hundreds. The first person to score ten points wins the game.

Sample round. Player A writes down the numeral 197, enters it in the calculator, and hands the calculator to player B with instructions to take away all the tens. Of course, B must recognize that the 9 is not just a 9 but that it represents 9 tens. B enters [−] [9] [0] [=]. The calculator displays 107, and B scores one point.

"Whole number buildup"

The rules for "whole number buildup" are exactly the same as for "teardown" except that players command each other to increase rather than decrease the original number.

Sample round. Player A writes down the numeral 4763, enters it on the calculator, and then hands the calculator to player B. A commands B to increase the 7 to a 9. B needs to recognize the 7 as 700 to be able to do this correctly—that is, to add 200 to the number. The correct result of 4963 gives the player one point. The first person to score ten points wins.

Decimal Numbers

Beginning in the fifth grade, students encounter the concept of decimal numbers. They begin to realize there is more to place value than ones, tens, hundreds, and so on. Fifth-grade teachers must help children understand these concepts. Those who use the standard techniques to teach these concepts should not forget that the calculator is ideal for reinforcing place value to the right of the decimal point.

"Decimal number teardown"

The game of "decimal number teardown" is similar to "whole number teardown."

Sample round. Player A writes a numeral that contains tenths and hundredths, for instance, 37.21. Player A hands the calculator to player B with the command to change the 2 to a 0. The equal sign may be pressed only once. B must recognize that the 2 is to the right of the decimal point and has a place value of two-tenths. B then enters [−] [·] [2] [=]. The calculator displays 37.01, and B scores one point. This exercise helps reinforce the concept that the 2 has a specific place value that depends on its location with respect to the decimal point.

"Decimal number buildup"

The "decimal number buildup" has rules similar to those for "decimal number teardown" except that the players increase rather than decrease the original number.

Sample round. Player A writes a decimal numeral—for example, 752.643—and enters it on the calculator before handing the calculator to player B. A commands B to increase the 4 to an 8. B must recognize that the 4 has a place value of four-hundredths. B adds .04, gets 752.683, and scores one point.

These activities with the calculator, designed to teach specific place-value concepts, are simple and fun. The teacher can easily design problems to teach the concept that fits the lesson of the day. Adding the calculator to the array of aids that are useful in teaching place-value concepts can facilitate the teaching of a sometimes difficult subject.

14

A Coordinate Graphing Microcomputer Unit for Elementary Grades

Betty Collis
Geoffrey Mason

USING microcomputers effectively in the elementary school can be frustrated by three common problems:

1. Often only one microcomputer is available for a class of students or even for the whole school.
2. Often the teacher has little or no programming skill and no more computer experience than an in-service workshop.
3. The mathematics curriculum seems overcrowded already, with little room for a new unit of study.

Here is a microcomputer project that addresses these problems. This instructional sequence, developed for a class of fourth-grade students, includes computer-literacy objectives with those of a regular mathematics unit. Classes were planned around the whole-class and individual use of one microcomputer, with the microcomputer itself being used in the context of the mathematics instruction. The computer materials included in the unit require only minimal computer experience on the part of the classroom teacher.

The mathematics topic for the unit was coordinate graphing, an important noncomputational aspect of mathematics that is often introduced in the fourth grade. The class that was chosen for the study consisted of twenty-four fourth-grade pupils (average age, nine years and two months). The unit consisted of thirteen forty-five-minute lessons given over a period of three weeks. A single Apple II Plus 48K microcomputer with disk drive, 12-inch monitor, and Silentype printer was mounted on a portable trolley and used throughout the lessons. Each lesson was conducted with one of the authors acting as teacher and the other as assistant, supervising the individual work of students at the microcomputer. An interested parent who is unacquainted with a microcomputer but who can use a typewriter could be trained for the task of assistant in about an hour.

The objectives of the unit fall into the two categories of coordinate graphing and computer literacy:

Coordinate Graphing Objectives

1. To describe a location by reference to horizontal and vertical alignment
2. To match ordered-pair notation with horizontal and vertical alignment
3. To name the Cartesian coordinates of any point located in the first quadrant
4. To graph a point in the first quadrant, given the coordinates
5. To give the coordinates of any point on the axes adjacent to the first quadrant
6. To graph a point on the axes, given its coordinates
7. To give the meaning and spelling of the terms *coordinate, first coordinate, second coordinate, horizontal axis, vertical axis, axes, origin, coordinate plane,* and *symmetry*
8. To experiment with concepts of symmetry and translation using figures in the first quadrant

Computer Literacy Objectives

1. To approach microcomputers with an attitude of enthusiasm and confidence
2. To recognize the meaning of the following terms as they relate to the Apple II Plus microcomputer: *microcomputer, monitor, input, program, floppy disk, disk drive, cursor, graphics mode*
3. To turn the Apple on and off and activate the disk drive and printer
4. To load a program using a disk drive
5. To locate the RETURN key and understand that it must be pressed after each input entry
6. To enter a simple command through the keyboard, using both print and graphics mode
7. To translate a sketch done in the first quadrant of the coordinate plane into a computer display of the same sketch
8. To load and run text and graphics files

These objectives were coordinated in a final project in which the students were required to create their own sketches in the first quadrant, enter the coordinates appropriately into the microcomputer set in the graphics mode, save their sketch programs on a floppy disk, and later load and run their programs for class discussion and display.

The three lesson plans that follow (1, 5, and 8) have been selected to demonstrate how the comptuer-literacy objectives were integrated in different ways with the coordinate graphing concepts. Also included are some materials used to evaluate the unit and a discussion of the results.

Lesson 1

The first day began with a computer vocabulary pretest (fig. 14.1). The items were read to the students with the option of "do not know" being clearly indicated as a possible choice. Results of this pretest showed that the children were quite unfamiliar with computer terminology. Their enthusiasm, however, was very high.

The class then formed a group around the Apple to watch the loading of the program "Hurkle" (available on "Elementary Disk, Version 1," from MECC, 2520 Broadway Dr., St. Paul, MN 55113). Words such as *microcomputer, disk drive, monitor, keyboard, input, output,* and *floppy disk* were introduced and then used throughout the entire unit with consistency and accuracy but without formal attempts at definition. Options 1 and 2 of "Hurkle" were played in the whole-class setting. These options involve guessing a hidden number on first a horizontal and then a vertical number line. The mathematical objectives of this activity were to develop the concept of describing location with a coordinate and to establish horizontal and vertical references. The microcomputer objectives included experiences in keyboard entry and the use of the RETURN key after completing each entry.

As the children worked at their seats on describing locations, two students at a time played both horizontal and vertical "Hurkle" with the microcomputer under the supervision of the assistant. In this way, all the students had hands-on experience with the Apple during the first two lessons.

The class seating plan was used to extend the description of location to a two-dimensional basis. Each student's desk was identified by coordinates formed by using row and seat numbers. Students responded to game-like directions such as these:

"Will the student in location (2,1) stand up."

"If your ordered pair is (3,4), raise your right hand."

"If your coordinates are (1,5), wave your hands."

"If your second coordinate is larger than your first coordinate, raise both hands."

"If the sum of your coordinates is 5, stand up."

Microcomputer Vocabulary Test

Draw a circle around the letter of your answer like this: (d)

1. Which one of the following is the most powerful?
   a) computer
   b) minicomputer
   c) microcomputer
   d) calculator
   e) don't know

2. The Apple is a
   a) computer
   b) minicomputer
   c) microcomputer
   d) calculator
   e) don't know

3. The monitor is what you
   a) press
   b) listen to
   c) switch on
   d) look at
   e) don't know

4. What you type on the keyboard of the Apple is called the
   a) input
   b) output
   c) cursor
   d) monitor
   e) don't know

5. A floppy disk looks like a
   a) soft cricket ball
   b) small phonograph record
   c) milk bottle top
   d) Frisbee
   e) don't know

6. Programs can be stored on
   a) floppy disks
   b) cassettes
   c) both floppy disks and cassettes
   d) monitors
   e) don't know

7. A disk drive is used to
   a) switch the Apple on
   b) make the Apple work
   c) show a picture on the Apple's screen
   d) feed a program into the Apple
   e) don't know

8. The cursor is
   a) a flashing signal
   b) the return key on the Apple
   c) the printed instruction sheet
   d) the TV screen on the Apple
   e) don't know

9. The graphics mode shows
   a) lines of writing
   b) pictures
   c) a flashing signal
   d) words in color
   e) don't know

10. A copy of what is showing on the TV screen can be obtained by using the
    a) disk drive
    b) monitor
    c) cursor
    d) printer
    e) don't know

Fig. 14.1

The grid format was introduced with an overhead projector and then extended to a map-reading application. Points with coordinates of the form $(0,y)$ and $(x,0)$ were graphed. The origin was labeled. A homework sheet (fig. 14.2) consolidated the day's concepts.

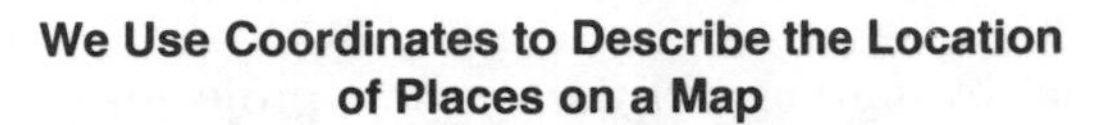

We Use Coordinates to Describe the Location of Places on a Map

Pretend this is a map of Victoria.

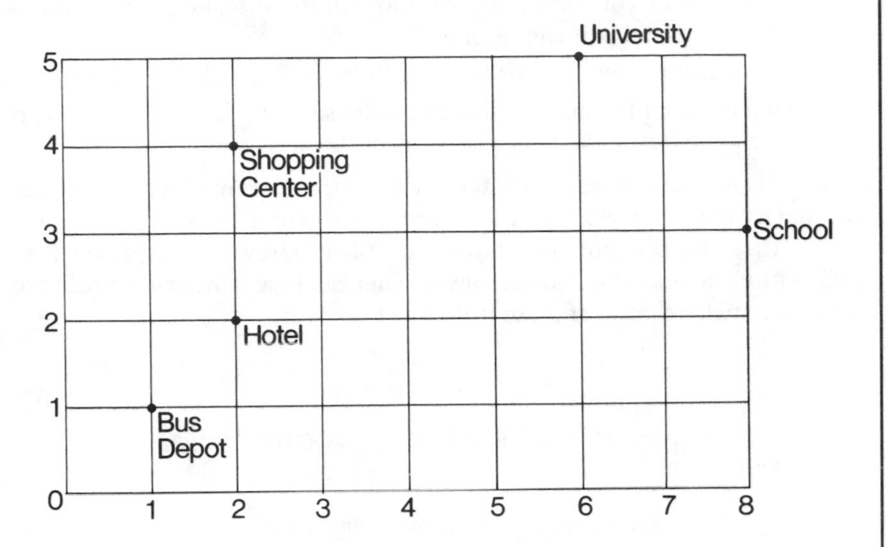

1. What ordered pair tells the location of
   - the University? (,)
   - the School? (,)
   - the Shopping Center? (,)
   - the Hotel? (,)
   - the Bus Depot? (,)
2. Your friend lives at (7,4). Find this location on your map, put a dot there, and label the point "Friend's House."
3. The coordinates of the location of your dentist's office are (3,1). Mark this point on your map and label it "Dentist."

Fig. 14.2

Lesson 5

Prior to lesson 5, individual students practiced turning on the computer and loading by means of a floppy disk another version of "Hurkle," which involved guessing the coordinates of a secret location in the first quadrant.

The students also worked at developing their own graph sketches in the first quadrant.

Lesson 5 began by showing with the overhead projector some special rules for drawing the next coordinate plane sketches:

1. Each "corner" or vertex of your graph must be on an intersection so that its ordered pair can be stated.
2. The starting point on your graph is to be marked with the letter *S*.
3. Your coordinates are to be listed in order so that you do not need to lift your pencil from the paper to connect them.

Introducing these rules at this stage enabled the students to combine their coordinate graphing work with their computer experiences through a program written for the unit. This program, called "Draw" in Applesoft BASIC, is listed in figure 14.3 so that any teacher can type it into an Apple, save it on an initialized disk, and use it in the classroom.

The "Draw" Program

```
10 REM DRAW A PICTURE BY BETTY COLLIS OCTOBER 1982
11 HOME
12 VTAB (8): HTAB (12): PRINT "*********"
13 VTAB (10): HTAB (13): PRINT "GRAPH A PICTURE"
14 VTAB (12): HTAB (12): PRINT "*********"
17 PRINT : PRINT : PRINT
20 INPUT "PRESS 'RETURN' IF YOU ARE READY TO GRAPH YOUR PIC-
   TURE . . ."; 8$
30 HGR
40 HCOLOR = 6
50 VTAB (21)
60 PRINT "TYPE THE COORDINATES OF YOUR CORNERS"
70 INPUT A,B
80 X = A * 20
90 Y = 140 - 10 * B
100 HPLOT X, Y
110 INPUT C, D
120 X = C * 20
130 Y = 140 - 10 * D
140 HPLOT TO X,Y
150 VTAB (23)
160 INPUT "DO YOU HAVE MORE POINTS ?"; Q$
170 IF Q$ = "NO" GOTO 200
180 PRINT : PRINT
190 GOTO 110
200 PRINT : PRINT
210 VTAB (21)
```

```
220 PRINT "TYPE THE NAME OF YOUR PICTURE AND YOUR NAME."
230 PRINT
240 INPUT N$
245 VTAB (4)
250 END
```

Fig. 14.3

In the whole-class setting, the "Draw" program was loaded and run using the coordinates of a graph that had been already drawn and displayed on the overhead projector. The program asks for the input of coordinates in sequence. After each ordered pair is entered, the program, using the high-resolution graphics ability of the Apple, connects the location of that entry with the location of the previous one. The children can thus see the graph being drawn as each pair of coordinates is entered. After each entry, the program asks if there is another pair of coordinates to enter. This input is included in the program to allow for maximum student-program interaction on a simple scale. When all ordered pairs are entered, the program asks the child to enter his or her name and a title for his or her graph. After this demonstration, each child had an opportunity to load and run the program using the simple graph shown in figure 14.4.

Triangle for "Draw" Program

Instructions:

After you have turned on the Apple and used the PR#6 instruction, type RUN DRAW.

Use this figure as you follow the instructions in the program.

A triangle by ______________________
(your name)

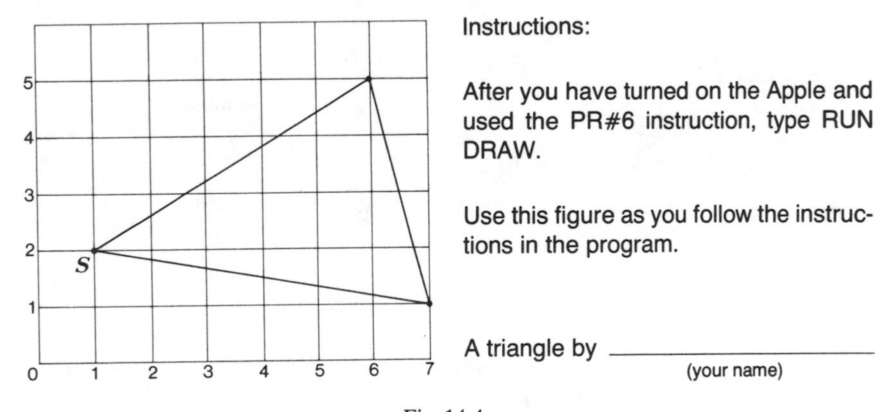

Fig. 14.4

Classwork and homework included *(a)* practice in drawing a graph by connecting ordered pairs in sequence and *(b)* practice in using the opposite approach of listing ordered pairs in sequence, given a sketch in the first quadrant and an indicated starting point.

Lesson 8

Lesson 8 began a procedure that continued through lesson 10—the final project for the children of making, saving, and running their own files. For this project each student prepared a graph sketch in the first quadrant following the procedure outlined in lesson 5. During these three lessons, each student, with a minimum amount of supervision, performed the following steps:

1. Insert the floppy disk with the "Draw" program.
2. Turn on the monitor and microcomputer.
3. Type RUN DRAW.
4. Follow the instructions of the "Draw" program by entering in sequence the coordinates of the vertices of your graph, a title for the graph, and your name.
5. Call for the instructor.

The instructor inserted an initialized master-class disk and saved the monitor display by typing in the following BASIC instructions:

BSAVE (child's name)'s GRAPH, A8192, L8192 (RETURN)
BSAVE (child's name)'s TITLE, A1024, L1024 (RETURN)

Figure 14.5 shows a monitor display of two of these files.

Fig. 14.5. Two examples of students' files

Students were then shown the names of their files in the catalog of the master disk. They were also given an experience with the printer. Each student was asked to choose a normal printout of white on black or an INVERSE display of black on white. The student activated the Silentype printer by typing PR#1 (RETURN), and the assistant then performed the rest of the operation.

While individual students were engaged in this computer activity, the rest of the class was consolidating the work of the unit by completing a workbook prepared by the instructors. They reviewed coordinate graphing topics, such as symmetry and translation, as well as various map-reading applications and microcomputer-related activities. (Fig. 14.6 is an example.) Copies of this workbook as well as materials and lesson plans for the unit are available from the authors (Department of Psychological Foundations in Education, University of Victoria, P.O. Box 1700, Victoria, B.C. V8W 2Y2).

Using the Apple Microcomputer

The steps for running a program on the Apple are listed below, but in *scrambled* order. Rewrite them so that they are listed in the *correct* order.

Scrambled steps:

Close the door of the disk drive.
Turn on the Apple.
Press the "RETURN" key.
Open the door of the disk drive.
Turn on the Monitor.
Type "RUN DRAW" and press "RETURN."
Slide in the floppy disk.

Fig. 14.6

Results of the Unit

Lesson 12 evaluated the students' work on the unit. Testing measures were a four-word spelling list, a test of the same ten computer-literacy items that had been used in the pretest (fig. 14.1), a ten-item hardware-identification test (fig. 14.7), and a paper-and-pencil test of the coordinate graphing topics. A performance test, which required students to operate the microcomputer, load and display their own graph and title files, and obtain a printout of the monitor display, was also given. Figure 14.8 shows the checklist used to evaluate this procedure.

Only two of the twenty-four students required prompting to remember the sequence of steps, and none needed help in the actual operation. Similarly, only two students failed to obtain 100 percent on the hardware-identification test—both as a result of reversing two keys. Table 14.1 gives the pretest and posttest item difficulties for the computer-literacy terms.

The lack of a control group and a lack of knowledge of the test-item difficulties for a larger group make it difficult to evaluate the coordinate graphing work. However, we are convinced that the students' enthusiasm for their individual projects did indeed facilitate their learning of these concepts.

Hardware Identification Test

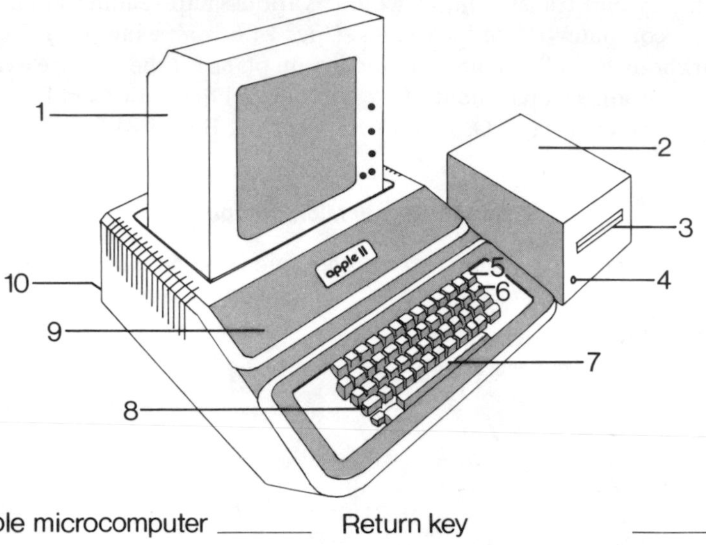

Apple microcomputer	______	Return key	______
Reset key	______	Floppy disk door	______
Disk drive	______	Computer on-off switch	______
Space bar	______	Shift key	______
Monitor	______	Red light	______

Fig. 14.7

Checklist for

Performance at the Computer

Performs without help each of the following steps:

1. Switches on the monitor ______
2. Lifts door of disk drive ______
3. Picks up floppy disk correctly ______
4. Inserts floppy disk, pressing it home ______
5. Closes door of disk drive ______
6. Boots DOS by switching on the microcomputer ______
7. Obtains High Resolution Graphics mode by HGR ______
8. Loads own program by
   BLOAD ______(Name)______'s GRAPH ______
9. BLOAD ______(Name)______'s TITLE ______
10. Turns on printer by PR#1 ______

Fig. 14.8

TABLE 14.1
Results of Microcomputer Vocabulary Test

Item Number	Percent Correct	
	Pretest	Posttest
1	33	71
2	4	83
3	50	88
4	46	71
5	38	100
6	38	79
7	25	92
8	8	83
9	21	46
10	17	63

Discussion

Although a steady supply of educational microcomputer software is now becoming available, very little of the material attempts to present an entire unit that integrates the microcomputer and software with an instructional sequence. The unit described here addresses this problem within the context of the three practical considerations mentioned earlier:

1. Only one microcomputer is available.
2. The classroom teacher has minimal computer expertise.
3. Objectives involving computer literacy are to be covered at the same time as a traditional unit of the curriculum rather than needing a separate (and perhaps unobtainable) block of free time.

The authors enthusiastically support the third recommendation of the NCTM's *Agenda for Action*—that "mathematics programs take full advantage of the power of . . . computers at all grade levels." We feel in particular that the early elementary grades offer much potential for the type of unit outlined in this report. One of our initial concerns was whether the levels of cognitive and psychomotor development of nine-year-olds were sufficient for learning microcomputer skills without undue frustration. This concern was dispelled in the first lesson. The children demonstrated great enthusiasm and self-confidence from the outset in handling the microcomputer, possibly because of experiences (now common to children) with computerized toys. However, because they were not familiar with the typewriter keyboard, some time was lost in the early stages of the unit while they searched for the correct keys. To remedy this, we suggest ten to fifteen minutes of whole-class focus on the QWERTY keyboard before attempting such a unit.

Our intent was to teach much of the microcomputer vocabulary and many of the skills "incidentally." This approach was completely successful for teaching the operation of the microcomputer and for naming the hardware.

The posttest mean computer vocabulary score of 77 percent, as compared to the pretest mean score of 28 percent, was considered to be quite satisfactory.

There is no question that the first of our computer-literacy objectives—to approach microcomputers with an attitude of enthusiasm and confidence—was achieved. All students eagerly attempted and successfully passed the performance test at the end of the unit. In addition, many unsolicited comments from the parents of the students, from other teachers, and from students themselves indicated a most positive reaction to the coordinate-graphing/microcomputer experience.

15

Using Computers in Teaching Mathematics

Marjorie A. Fitting

JUST as it is unthinkable in today's world to operate a store without a cash register or an office without a typewriter, it is fast becoming unthinkable to operate a business without a computer. Businesses use computers to keep records, to discover trends, and to predict future markets. Mathematics curricula, too, must take full advantage of the power of computers at all grade levels.

In the teaching of mathematics, computers can give students a great diversity of experiences. They can provide individually managed drill and practice, tutorials, classroom or individual demonstrations of concepts, and simulations of experiences. In addition, computers are tools for the teacher or student to use in problem-solving experiences.

Unfortunately, computers can also be misused in the classroom. Having to read a paragraph on a computer and answer several multiple-choice questions would be just as boring to students as having to work at the same task from a book. Games, however motivational, may not have educational merit. Drill and practice without immediate feedback can be done just as well with pencil and paper. Watching the computer without any interaction other than occasionally pushing a carriage return is not active learning. Computers should be used for what they do uniquely well and not for duplicating what can be done just as well in other ways.

Drill

The computer is a tireless drillmaster that can present the facts to be memorized over and over again to each individual student until mastery is achieved. Material ranging from addition tables to trigonometric identities appears on the screen, and the computer immediately checks the student's response. In well-written drill programs, a variety of comments that include both positive and negative feedback motivate the student to memorize the information. The computer can also regulate the time available for response and keep records of each student's progress.

Practice

Although a drill program is appropriate when material is to be memorized, a practice program is more appropriate when general principles or algorithms are to be applied to specific situations. Again, the computer responds with a variety of positive comments when the work is correct; when it is incorrect, the computer branches to subroutines that help the student review the principles that are to be applied. In a practice program for rounding numbers to the nearest ten, the computer may box the units digit when a mistake is made in rounding up instead of down and ask, "Is the units digit less than 5?" When the student rounds to hundreds instead of tens, the computer may ask, "Which digit is the tens digit?" The ability to handle common mistakes by reviewing basic principles is an attribute of a good practice program.

Computer-managed Drill and Practice

Many commercial companies have developed computer-managed (CMI) drill-and-practice programs in arithmetic. The SRA (**17**)* programs for the Atari (**3**) 800 computer, for example, supply the teacher with diskettes that can be used with an Atari computer configured with a single disk drive. One diskette includes the drill-and-practice programs, and a second contains the students' data files, which place students in appropriate lessons and record their progress. A third diskette attends to management tasks: entering the students' names, grades, and computer passwords, specifying their placement into levels of work, and developing reports on their progress. The student begins by using the student data file diskette to tell the computer who is going to be using the programs. The program on the data diskette assigns the student to a lesson and asks the student to switch to the drill-and-practice program diskette. After the lesson is completed, the student replaces the data diskette to record the progress made. At the end of the day, the teacher uses the teacher diskette to produce a report on student progress for that day.

An attractive feature of some CMI programs is that they can provide diagnostic testing, tutorials in areas needing improvement, drill and practice, and achievement testing. Furthermore, the computer can keep full records

---

**Note:* Bold numbers in parentheses throughout this article refer to notes at the end of the article listing sources of computers and programs.

on each student and prepare reports for the teacher either on individuals or on the entire class. To use such a program effectively, the classroom must be equipped with either a terminal connected by telephone line to a time-sharing computer system or a microcomputer system that includes sufficient storage media. Some microcomputer systems use a connecting network such as that designed by Nestar Systems (**14**), which allows many microcomputers to be connected to a common disk drive and printer and allows up to thirty-two students to work simultaneously with the same set of programs.

One complete curriculum in arithmetic for grades K–8 using the TRS-80 (**19**) Model I, Level 2 computers, was developed by the Dallas Independent School District (**7**). This curriculum has the strength of having been developed and used successfully on paper first, so that the educational soundness of the material was established and the accompanying teacher role was spelled out. The programs were then written so the computer would take over the repetitive functions of testing, drill, practice, and record keeping while teacher instruction continues in the classroom.

Classroom Assistance

Demonstrating algorithms on a large video screen helps make a classroom presentation more efficient. The teacher can have each step programmed so that the next will be displayed when the return key is pressed. The teacher is free to answer questions and help individuals while the computer does the work. After the presentation is over, a student can review the material by rerunning the program.

An especially effective use of this technique was developed by Helen Mah (**13**) on the Compucolor II for a linear algebra presentation on row reduction. The matrix that is to be reduced is entered in the standard matrix format. The computer lists the three elementary row operations and asks such questions as these:

- Which operation do you wish to use?
- Which row is to be changed?
- Which row is to be used to change it?
- By which factor is row 3 first multiplied?

Each row operation is performed with time delays between each entry and in a new color so that the change can be easily observed. Since many choices of ways to proceed in row operations are available, this procedure allows the students to participate in choosing their next steps and helps them find errors in work they have previously done on paper.

In a classroom demonstration to develop the concept of least common multiple, a computer program can generate long lists of multiples of each of the numbers being considered. The students can then pick out the common multiples and the least common multiple. Without the computer, the teacher and students might tire of finding multiples before the students have fully understood the concept that the set of multiples, indeed of common multiples, is infinite and that we seek only the least one.

Programs that demonstrate algorithms can free the teacher from the blackboard, perform the routine arithmetic calculations correctly, and display many more examples than might be possible otherwise. Further, students can review the algorithm or see lessons they have missed.

Tutorials

The purpose of a tutorial program is to present material to the student in an interactive experience. Such a tutorial can be used as a student's first exposure to a topic, but it is more likely to be used as a follow-up or review lesson after the topic has been studied in the classroom. Different tutorials on the same topic can provide different approaches, thereby allowing students to choose the most effective method for their learning capabilities or habits. For instance, a presentation of the algorithm for long division could provide a stairstep development in which each new step is followed by practice and a review of all steps up to that point. A global approach might display a flowchart of the entire algorithm that is highlighted as each step is illustrated and highlighted again when each step is needed during practice. In either of these approaches, the program can refuse to accept a wrong answer at any stage and can explain what the right answer should be so that the student's efforts are directed correctly.

In a global and spiral approach to the concept of slope (algebra in grade 9), Howard Jensen's (**12**) program first displays sloping lines and describes each as having negative slope or positive slope. It then describes slope in terms of a roof (its rise and run, for instance) as a real-life application. It next applies coordinate axes to the line and shows how the rise and run are measured in terms of coordinates. Finally, it presents the formula for finding the slope. To answer students' questions about slope, the computer asks them to enter points; then the computer computes the slope and shows its work. One advantage of a computer presentation is that students can be asked questions at each stage and their responses can be corrected immediately before they proceed. Another advantage is that they can proceed at their own rate and can be given additional examples at each level of development. At the end of each tutorial, the computer will also patiently provide practice, correct errors, and give praise.

Tutorials can also be developed using an analytic approach or an applications approach. In the analytic approach, students are helped to analyze what is needed if they are to get where they are going. This approach is particularly applicable to developing techniques of proof in geometry. For example, if a student wished to prove that the angle bisector is equidistant from the sides, the computer would focus the student's attention on what is to be proved: first, what is meant by *equisdistant,* and second, how the triangles formed by the line segments measuring the distance can be proved congruent. In the applications approach, the student learns to relate a real-life situation to the mathematical model. For example, in division using the set partition approach, the student is asked to divide twenty-four pieces of apple among six people.

A discovery approach to the concept of the slope of a line might involve asking the student to choose values for m in the equation $y = mx$. The computer then graphs each line on the coordinate axes. After students see that the value of m affects the slope, they might be encouraged to try to find values for m that will cause the line to go through specific sets of points such as (1,1), (2,1) (3,1); $(^{-}1,1)$, $(^{-}2,1)$, $(^{-}3,1)$; (2,1), (4,2), (6,3); and so on. Successive levels of difficulty could include predicting which points the graph of $y = 2x$ passes through, and finding specific m values to make $y = mx + 5$ go through other specific points. Using the computer encourages students to make hypotheses and to check them because it takes very little computational effort and the check is provided quickly. The computer forgives poor guesses; when the next guess is made, the old guess is erased. Students are often more willing to guess and check and are not so afraid to make mistakes (and learn from them) when there is no permanent record. Furthermore, students often understand better the concepts they have "discovered" themselves and often retain them longer.

In a tutorial involving slope, some analysis could be combined with a map-reading application (fig. 15.1).

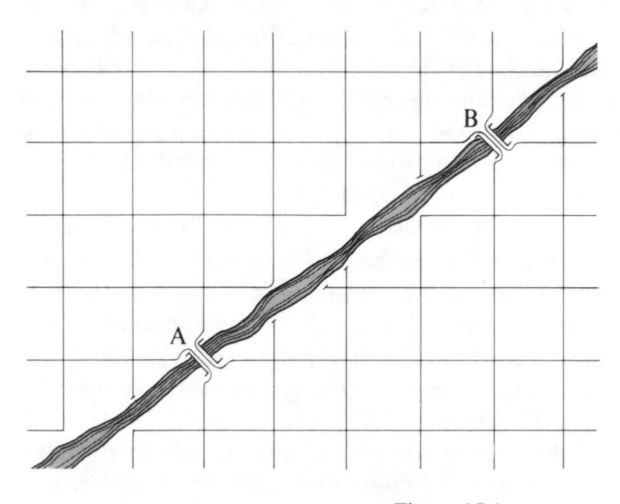

Tell a fisherman how to get from bridge A to bridge B.

Figure 15.1

As the student responds to the instruction, the computer will supply more information as needed; "I only know the words UP and RIGHT" and "I didn't find the bridge, so I came back." Each instruction will move the fisherman, but incorrect ones will return him to bridge A. With each example, of course, the bridges can be relocated. The rules can change later so that the instruction for UP 3, RIGHT 4, must be written as 3/4.

Tutorials have the advantage of providing alternative approaches to topics, allowing the students to work at their own pace and incorporating practice and correction at each step of the learning sequence. They cannot, however, replace the teacher, since no program can foresee all the questions students may have or the diverse backgrounds students bring to the classroom. In Howard Jensen's program on slopes, if a student makes three consecutive errors, the computer suggests that the student see the teacher for help.

Demonstrations

The discovery approach provides students with many examples of a concept and gives each student the opportunity to generalize from the examples and to predict from that generalization what will occur in additional examples. This approach is often forsaken in the classroom because it requires more time than a deductive or an authoritarian approach. The computer can provide many examples in a short period of time and restore the joy of discovery to the student.

In demonstrations, the student supplies the variables and the computer displays the results. In a demonstration on the sampler diskette that Apple Computer, Inc. (**2**) supplies with each disk drive, a third-degree polynomial is graphed and the user is asked what changes are desired in the coefficients. The computer graphs each new polynomial individually, thereby allowing the user to discover the effect of changing coefficients. For the discoveries to be made, however, the user may need some guidance. An accompanying instruction sheet could make such suggestions as, "Change only the coefficient of x. How does this coefficient change the graph?"

In a demonstration developed for the Compucolor II (**5, 10**), the function Y = A * SIN (B * (X + C)) is demonstrated. Students vary A, then B, and then C to discover the effects of each. During a variation of A, for example, each graph appears in a different color so that each may be compared, and the equations appear on the screen in the corresponding color. After each variable is demonstrated separately, the student may vary the three variables simultaneously to discover interaction.

Using the discovery method helps students assimilate what they have discovered. The computer provides the many examples, the patience, and the applications so that the student is assured of making the discovery.

Simulations

> You have been chosen to rule ancient Sumeria [*sic*]. You will make decisions on how much land to buy or sell, how much wheat to feed to your people, and how much land is to be planted each year. Each person requires 20 bushels of wheat per year to live and 1/2 bushel of wheat is required to plant one acre. May you rule happily and in good health.

Rich in possible arithmetic computations (and some strategy when the wheat supply runs low), this computer simulation, known as "Hammurabi," involves students in decision making. Chance plays a role, for the wheat harvest can be low or high and plague can beset the people. However, if the students' arithmetic is correct and their decisions are wise, they can rule successfully for many computer years.

Simulations involve the students in active learning, problem solving, and decision making in a way unequaled by any other medium. Students do computations and make predictions of how things will come out if they make a certain decision. Then they verify their prediction by making the decision and seeing the consequences.

Another interesting simulation appropriate for the mathematics classroom is "Stock Market Game" (**15**). Many versions have been written, some

Brian McClory watches intently as his sister, Carrie, creates a picture using LOGO.

of which simulate very closely the fluctuations of the stock market. "Hammurabi" and "Stock Market Game" are generally available through computer-user groups. "Lemonade Stand," a business enterprise for elementary school children, is available on a diskette provided by Apple Computer, Inc. (**2**). It is unfortunate that few simulations of this type are available, since they encourage students to apply their arithmetic skills.

A simulation developed at Massachusetts Institute of Technology has students use the keyboard of a computer to direct and move about a turtle (robot) or a simulated turtle (small triangle on the video screen). The student can instruct the turtle to raise its pen (PENUP) or lower it (PENDOWN); the "pendown" position leaves a trace of the turtle's wanderings on paper or on the screen. The computer language LOGO, available for both the Apple and the TI 99/4 (**18**), developed from these efforts. LOGO is both a simple language that can help elementary school students learn about programming and logical thinking and a sophisticated language that can be useful in teaching the fundamentals of computing at the college level. LOGO can be used interactively with the computer, putting the student in control of the computer with a minimum amount of knowledge and with immediate results. The graphic applications of LOGO include "Turtle Geometry" (**1**), a medium for exploring the world of geometry by discovering relationships that also suggest methods of proof. LOGO geometry is an integration of the computer and the discipline, not just another medium used to demonstrate concepts.

Problem Solving

Problem solving often requires collecting and analyzing data, making a hypothesis, and checking the hypothesis. Computers are an excellent tool for helping collect the data.

On the elementary level, the game "guess my number" involves students in collecting data about an unknown number. They decide which numbers are greater and which are lesser and try to develop a strategy that will enable them to guess the number in the fewest number of guesses. The program for playing this game on the computer takes few steps and allows students to develop their own game strategy.

Students in both elementary and junior high schools study mathematical patterns. Computers can generate patterns and provide students with requested data so they can guess the pattern and test their guesses. Programs to generate arithmetic sequences, square numbers, triangular numbers, Fibonnaci numbers, and geometric sequences are in most software libraries.

In advanced mathematics classes at the high school level, students learn to find real roots of polynomials and limits of functions. Because of the complex arithmetic computations required, very few examples are done arithmetically in the classroom. Even pressing the buttons on a calculator for these activities can be exhausting! With even very short computer programs, however, insight into the technique of successive approximations can be acquired and many examples can be worked.

Students run the program in figure 15.2, for instance, to find the limit of sin $(x)/x$ as x approaches 0. They can run such programs for values of x approaching 0 and acquire a substantial amount of data regarding the behavior of the function in this domain. The program in figure 15.2 can also be used to evaluate any polynomial by substituting the expression for the polynomial; for example, Y = X * X − 3 * X + 13, in step 20. Since it is easy to acquire data, students do not jump to conclusions prematurely. Furthermore, they can keep records of the data as it is generated and make their own analysis without having merely to accept what the authority says.

```
Program (in BASIC):
10 INPUT "ENTER X ";X
20 Y = SIN (X) / X
30 PRINT X,Y
40 GOTO 10
```

Fig. 15.2

It should be clear that by using calculators in problem solving students can focus on the strategy of problem solving rather than on the computation itself. When the arithmetic computations are more complex, the computer can serve as a tool.

Programming and Other Computer Skills

In Santa Clara County, California, students in grades 4–8 are often offered instruction in computer literacy, which includes becoming familiar with what computers can do and using computer programs created by others. In addition, each high school offers at least a one-semester course in programming in BASIC, with many high schools offering more advanced courses as well. A week or two of programming in BASIC is also incorporated into such mathematics courses as algebra and general mathematics. Some high schools also offer instruction in programming in PASCAL. The burden for teaching these courses falls on the mathematics teachers.

Many high school textbooks are incorporating flowcharts into instruction. Flowcharts are useful for displaying the steps of an algorithm as well as for outlining any order of procedures, such as the order in which activities are to be done in a third-grade class. Even if a school cannot afford to purchase a small microcomputer (list prices on the TI 99/4A (**18**) begin at $199), the concepts of flowcharts can be incorporated into instruction.

Games

We are all more motivated when drill, practice, and demonstrations are incorporated into games. In the Cupertino Union School District, Bobby Goodson (**11**) uses games with underlying educational objectives to introduce elementary school students to the computer and its keyboard. Games can help students understand concepts as well as provide drill and practice. Bill Finzer at San Francisco State University (**8**) developed a game on tape for the Commodore PET (**4**) computer in which a plane drops objects onto targets located on the number line. The player specifies a rational number by typing in the name of a fraction that corresponds to it. The plane orients itself over the specified point on the number line and drops the object. When it misses the target, the player can see whether the rational number was greater or less than the target and can try again. Success is rewarded by the target appearing to explode.

Another game with fractions, "Frog Race" (**9**), was developed on the Compucolor II (**5**) computer. Each frog is randomly assigned a fraction between 0/1 and 1/1, which tells the average length of its jump, and each player chooses a frog. The frogs hop across the screen, and the one having the greatest fraction wins the race. (One to three students may play, and each player has a chance to choose first.) In both these games, students develop a better concept of *greater than* and *less than* with fractions and rational numbers.

Summary

Using the computer for drill and practice, tutorials, demonstrations, and simulations requires a substantial investment in the development of programs. Because of this requirement, most teachers will search for available programs that do what they require; some schools may even buy a particular computer because the programs its teachers want are available on it. If the local university has a mathematics department active in program production, it might be wise to buy the same type of computer used there and to request that particular program be developed by students or faculty from the university. Morgan Hill (Calif.) Unified School District bought computers for the Live Oak High School to match those of the elementary schools so that high school student programmers could develop programs for the elementary level.

Unfortunately, most of the commercially available programs are of the drill-and-practice variety, and little programming has been done with discovery-oriented materials. Microcomputers lend themselves especially well to graphic presentations, and although the development of graphics takes a longer time, the results seem well worth the effort. County educational distribution agencies can serve a valuable function by becoming a collection-and-distribution agency for teacher-developed programs and by displaying commercially available programs for review before purchase. One example of such a center is the San Mateo Microcomputer Center (**16**), which also houses the CUE (**6**) software library.

The use of the computer for problem solving, however, is not limited by the constraints of needing extensive software. By learning a minimum number of programming techniques, students can make use of the power of computers to analyze data, develop generalizations, and apply techniques for problem solving. Even when students don't know anything about programming, teachers can use computers for problem-solving instruction by making short programs available to generate specific data.

REFERENCE NOTES

1. Harold Abelson and Andrea A. diSessa, *Turtle Geometry*, Cambridge, Mass.: MIT Press, 1980.
2. Apple Computer Inc., 10260 Bandley Dr., Cupertino, CA 95014.
3. Atari, 1265 Borregas Ave., Sunnyvale, CA 94086.
4. PET and VIC20 are products of Commodore Business Machines, 761 Fifth Ave., King of Prussia, PA 19406.
5. Compucolor Corporation, P.O. Box 569, Norcross, GA 30071.
6. CUE, Computer-Using Educators, c/o Don McKell, Independence High School, 1776 Independence Dr., San Jose, CA 95133.
7. Dallas Independent School District, 912 S. Ervay, Dallas, TX 75201.
8. Bill Finzer, Department of Mathematics, San Francisco State University, 1600 Holloway Ave., San Francisco, CA 94132.
9. Marjorie A. Fitting, "Frog Race," on Mathematics Diskette #1 for Compucolor II, in the CUE library deposited with the San Mateo Microcomputer Center.
10. Marjorie A. Fitting, "Sine Demonstrations," one of thirteen programs with text and diskette for Compucolor II, *Programs for Trigonometry*, Metier, P.O. Box 51204, San Jose, CA 95151.
11. Bobby Goodson, Computer Resource Teacher, Hyde Junior High School, 19325 Bollinger Rd., Cupertino Union School District, Cupertino, CA 95014.
12. Howard Jensen, "Beginning Slopes," one of a series of programs for first-year algebra developed at Pioneer High School, 1290 Blossom Hill Rd., San Jose, CA 95118.
13. Helen Mah and Marjorie A. Fitting, "Row Reduction," a program on Mathematics Diskette #7 for the Compucolor in the CUE Softswap (**16**) library.
14. Nestar Systems, 2585 E. Bayshore, Palo Alto, CA 94303.
15. Steve Norms, "Stock Market Game," *Creative Computing*, 1978, pp. 154–56. From a program developed by the Huntington Project.
16. Softswap, San Mateo County Office of Education, 333 Main St., Redwood City, CA 94063. This center gathers programs contributed to the Computer Using Educators library for the Apple II (13 sector diskette), Commodore PET, TRS-80, Atari, and Compucolor II computers. A directory of programs for each computer may be obtained by writing to the center. Diskettes are provided at cost of the diskette plus a small service charge for the copy service.
17. Science Research Associates, P.O. Box 10021, Palo Alto, CA 94303.
18. TI 99/4 is a trademark of Texas Instruments, P.O. Box 1444, Houston, TX 77001.
19. TRS-80 is a trademark of Tandy Corp., 400 Atrium, 1 Tandy Center, Fort Worth, TX 76102.

Recommendation 4

STRINGENT STANDARDS OF BOTH EFFECTIVENESS AND EFFICIENCY MUST BE APPLIED TO THE TEACHING OF MATHEMATICS

16

Favorite Days in the Classroom

Joan Duea
Earl Ockenga
John Tarr

STUDENTS in a sixth-grade classroom were huddled in twos. Each partnership was busy writing the word *free* in one square of a four-by-four grid and then filling in the other fifteen squares with numbers the teacher had written for display on the overhead projector. Fifteen problem cards were displayed on the chalkboard tray, and the projected numbers represented the answers to these problems. The students were preparing to play one of their favorite games, problem-solving bingo. Each team would get a calculator, choose a problem card, work together to solve the problem, find the answer on their grid, and mark the answer with an X. The object of the game

was to be the first partnership to get four answers in a line—horizontally, vertically, or diagonally.

Today would be a favorite day in this sixth-grade classroom. Learning would take place effectively and efficiently because students were—

- actively and productively involved, sharing their thinking and making decisions together;
- choosing for themselves which problems they would solve;
- working enthusiastically, motivated by the competition of the game.

Students and teachers deserve many favorite days in the classroom. As you reflect on your own favorite classroom days, you may find that they have three common characteristics: (1) the dynamics of the classroom were positive, with student discussion and involvement contributing to the learning goal; (2) the use of media was efficient, with the students' visual, auditory, and tactile senses focused on the learning; (3) the instructional activity was an effective way to meet the lesson's goal.

Classroom Dynamics

Dynamics, good or bad, are present in every classroom during every period of the school day. The interaction among pupils and between pupil and teacher can be planned to help pupils get the most from the learning experience. As teachers make lesson plans, they must give as much consideration to the dynamics of the classroom as they do to objectives, concepts, and materials. Will the students learn best by working alone, with a partner, in teams of four, or as a member of the total group? Will the students depend on the teacher as the learning resource, or will they learn from each other?

Case study: The elevator problem

Students in a fourth-grade classroom were placed in teams of four. Each team received the following problem and was challenged to use its team resources to solve it:

> If an elevator has a capacity of 2000 pounds, how many people will it hold? As the class worked, the teacher moved from team to team listening to the conversation and evaluating student understandings.

The teams found that there was no immediate and obvious way to solve the problem. They realized that they needed to collect their own data and use this information in solving the problem. One student assumed the elevator would be filled with students and started collecting the weights of other teammates. Another remembered the weights of several professional football players and thought of filling the elevator with NFL players. Another student asked if anyone on the team knew how much their mother or father weighed. A fourth went to the teacher and asked, "How much do you weigh?"

This would be a favorite day in the classroom for these students and their teacher. The emotional climate was positive. There was an openness and a willingness to respect ideas. Everyone realized that there was no such thing as a "dumb" question or a "dumb" comment. The teacher was a facilitator and a model in this learning environment as well as the presenter of problem-solving experiences. This kind of environment stimulated thinking because—

- students were encouraged to share their thinking, to listen, and to react to their classmates' ideas;
- team members gained greater insight into problem-solving processes by verbalizing their ideas;
- team members became greater risk takers through their group effort, and thus their problem-solving confidence increased.

Media in the Classroom

Media can enhance the learning of mathematics. The effective and efficient use of overhead projectors, tape recorders, films, calculators, computers, and manipulatives can stimulate, enrich, and intensify learning.

Case study: What's my number?

Students in a ninth-grade general mathematics class were busy making a list of numbers that fit the clues projected on a screen. As students put their pencils down to indicate they had finished with one clue, the teacher would remove a plastic chip from the overhead display to reveal another clue. The students would check their list of numbers and eliminate those that no longer fit the conditions specified. The object of the game was to find what number was being described.

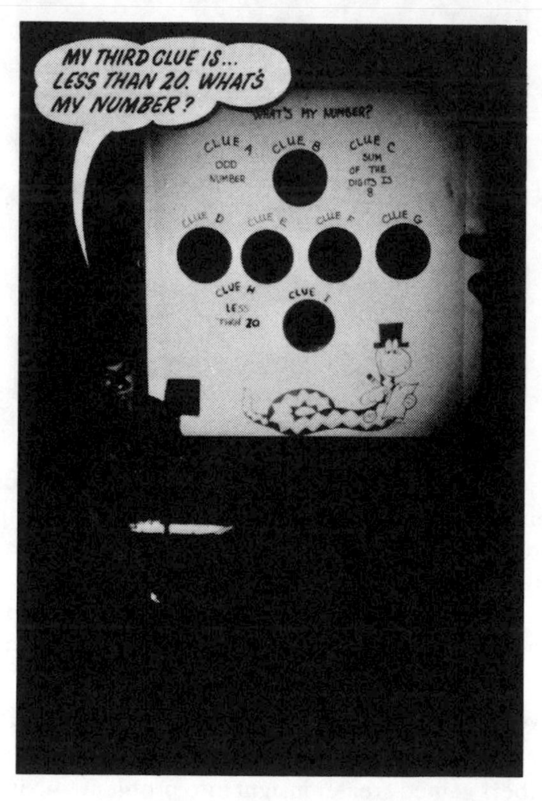

Today would be a favorite day in their classroom. The use of media—in this instance, the overhead projector—made learning both efficient and effective because—

- the activity could be placed on a transparency before class time, thus saving instructional time;
- the students' attention is focused on the projected visual;
- the teacher could control the pace of the activity by revealing a small portion of a transparency at a time.

Case study: Find the greatest product

Each student in a fifth-grade classroom had written the digits 1, 2, 3, 4, and 5 separately on five small pieces of paper. They were arranging them to make a three-digit number and a two-digit number, then using their calculators to find the product of the two numbers. They listed their factors and product on paper, then rearranged their slips of paper to form other three-digit and two-digit numbers. The object of the activity was to find the three-digit and two-digit factors that resulted in the greatest product.

The appropriate use of calculators in the classroom can lead to a favorite day for both students and teachers because—

- calculators allow students to compute accurately in a short period of time;
- calculators shift the focus from computational skills to problem-solving skills;
- calculators give students equal opportunities to participate by removing the computation barrier.

Learning can be exciting in the classroom. The technological advances of today's media, especially in the area of graphics and sound effects, can stimulate students and cause them to become totally absorbed. The appropriate use of media in our schools can help us attain greater intensity for learning.

Instructional Activities

What kind of activity capitalizes best on students' resources, uses media effectively, and most importantly, meets the objectives of the lesson? This is the question that effective teachers ask themselves as they plan each day's mathematics lesson.

Case study: How sure are you?

Students in a tenth-grade geometry class were solving problems and carefully checking their answers. As they completed each problem, they made a wager depending on how sure they were of their answer. If they were "very sure" of their answer, they could wager ten points; if they "thought" their answer was correct, they could wager five points; if they were "not sure," they could wager two points. Students received the points they wagered on correct answers and lost the points they wagered on incorrect answers. The object of this activity was to get students to reflect on their answers.

The activity made optimal use of the classroom period because students—

- took time to check their work;
- looked frequently for an alternative method of solving the problem to be sure their answer was correct;
- made judgments regarding the accuracy of their thinking.

Case study: "Solve and compare"

Students in a third-grade classroom were involved in the instructional activity called "solve and compare." Problem cards, scratch paper, and paper clips were on a table. Students were freely selecting a problem card to solve. They solved the problem, wrote their name and answer on a piece of scratch paper, and clipped it to the back of the card. Then they returned the card to the table and selected another. As the work progressed, they could compare their solutions to the attached answers of other students. When their answers differed, students consulted with each other. The total classroom period was productively used on problm solving.

"Solve and compare" is an activity that students ask to do again. Efficient and effective activities such as this one stimulate student learning as well as meet student needs.

- Students use each other as resources for comparing answers and resolving differences.
- Students improve their communication skills as they explain their solutions to classmates.
- Teachers watch and listen while students become the doers and talkers.

Every teacher enjoys sharing good days in the classroom. Not every day of the school year falls into this category—and probably cannot be expected to. However, good days for students and teachers will promote good learning. Effective and efficient teachers can use methods like these to inspire many favorite days in the classroom.

Recommendation 5

THE SUCCESS OF MATHEMATICS PROGRAMS AND STUDENT LEARNING MUST BE EVALUATED BY A WIDER RANGE OF MEASURES THAN CONVENTIONAL TESTING

17

Diagnosing Student Error Patterns

Mary Ann DeVincenzo-Gavioli

STANDARDIZED and teacher-made tests are traditionally used to measure students' proficiency in mathematics. Unfortunately, teachers can misinterpret students' written computations on these tests. The purposes of this article are (1) to demonstrate specific examples of how written computations can be misjudged and (2) to suggest techniques that can improve teachers' diagnoses of students' capabilities from written computations.

It should be noted that in conventional testing seemingly correct responses as well as incorrect ones often misrepresent a student's understanding of the problem. The application of an erroneous algorithm can sometimes result in a correct answer (DeVincenzo 1980). For example, in the problem $3/4 - 2/3$, using the mathematically incorrect procedure of subtracting the original numerators and finding a common denominator for the fractions will yield a correct answer! (See Examples 3 and 4.)

Will a specific incorrect answer direct a teacher to the underlying mathematical misconception? Not necessarily! Although the incorrect answers provided in multiple-choice tests are generally based on common errors and misconceptions (Epstein 1968, p. 319), a student's selection of a particular incorrect answer does not yield adequate diagnostic information about the nature of her or his difficulty in mathematics. This is because (1) the same incorrect answer may result from different misconceptions (see Example 5) and (2) correct computations may have accompanied such a selection. A student might select an incorrect answer on a multiple-choice test because of an inability to recognize the computed answer rather than an inability to compute. For example, one student was observed to compute $2/3 \div 4/5$ correctly as 10/12 and then to select 12/10 as the answer because "10/12 was not one of the choices, and I thought 10 and 12 had to be in the answer somewhere." The correct choice was 5/6. (Examples 4 and 5 demonstrate the importance of the "think aloud" diagnostic test.)

Consequently, the responses recorded for mathematics tests do not always accurately reflect students' strengths and weaknesses in a particular topic. The results of such tests *cannot* suggest appropriate remedial activities to the teacher. Specific diagnostic techniques that can expand on written computations and improve teachers' diagnostic skills are needed!

In the following examples suggested diagnostic techniques will be discussed in conjunction with some specific, though difficult-to-detect, misapplications of algorithms.

Example 1: The correct answers in figure 17.1 do not reveal a certain ninth-grade student's faulty algorithm. The faulty algorithm, always recording a 1 above a column when carrying in an addition problem, was uncovered when the student (whose computations are supplied in figure 17.2) explained her solution for the problem 89 + 95 + 39.

$$\begin{array}{r} 83 \\ +\ 75 \\ \hline 158 \end{array} \qquad \begin{array}{r} 82 \\ +\ 19 \\ \hline 101 \end{array}$$

Fig. 17.1. A ninth-grade student's computations for the addition of whole numbers

$$\begin{array}{r} {}^{1}89 \\ 95 \\ +\ 39 \\ \hline 213 \end{array}$$

Fig. 17.2. A ninth-grade student's computations for 89 + 95 + 39

Suggestion 1

Examining the solution to one example will not always reveal a student's error. Have students solve at least three examples of a particular type; a number of researchers (Ashlock 1976; Cox 1974; Graeber and Wallace 1977; Inskeep 1978; Pincus et al., 1975; West 1971) have suggested at least three examples of a particular type to facilitate diagnosis.

Example 2: Responses to three whole-number subtraction problems (fig. 17.3) appear to indicate that this particular student lacked a knowledge of the subtraction algorithm. However, an examination of the computations that led to these answers, recovered from the student's scrap paper (see fig. 17.4), reveal a complete understanding of the subtraction algorithm with addition used as a check! After checking the problem, the student had recorded the final figure in the checking process as the answer! (See fig. 17.4.)

$$\begin{array}{r} 49 \\ -\ 3 \\ \hline 49 \end{array} \qquad \begin{array}{r} 875 \\ -\ 23 \\ \hline 875 \end{array} \qquad \begin{array}{r} 250 \\ -\ 22 \\ \hline 250 \end{array}$$

Fig. 17.3. A ninth-grade student's responses to whole-number subtraction problems

$$\begin{array}{r} 49 \\ -\ 3 \\ \hline 46 \\ +\ 3 \\ \hline 49 \end{array} \qquad \begin{array}{r} 875 \\ -\ 23 \\ \hline 852 \\ +\ 23 \\ \hline 875 \end{array} \qquad \begin{array}{r} 250 \\ -\ 22 \\ \hline 228 \\ +\ 22 \\ \hline 250 \end{array}$$

Fig. 17.4. The computations accompanying the solutions in figure 17.3

Suggestion 2

Design exercises for tests and homework that require students to record and submit as many computations as possible. Such exercises are a good alternative to multiple-choice exercises, for which the accompanying written computations are usually not required. When multiple-choice tests are administered, require the students' written computations to be submitted. They can give you an opportunity to analyze solutions for such error patterns as specific faulty algorithms used as solution techniques.

Example 3: A student's correct final answers for two of the three subtraction-of-fractions problems in figure 17.5 would lead a teacher to assess this student proficient in the subtraction of fractions and to interpret the one incorrect solution a result of carelessness. However, the student who computed the problems in figure 17.5 explained that in each instance the numerators were subtracted. This difference was placed in the numerator of the answer, and a common denominator was computed and used as the denominator in the final answer. Note that the faulty algorithm resulted in a correct answer for the majority of these problems. Such an occurrence, if not diagnosed and corrected, can help reinforce a student's confidence in incorrect techniques of solution. (Confidence in incorrect algorithmic techniques was reported by MacKay in 1975.)

$$\frac{4}{5} - \frac{2}{3} = \frac{2}{15} \qquad \frac{3}{4} - \frac{2}{3} = \frac{1}{12} \qquad \frac{2}{7} - \frac{1}{5} = \frac{1}{35}$$

Fig. 17.5. A ninth-grade student's solution for three subtraction-of-fractions problems

Suggestion 3

Since faulty algorithms can sometimes produce correct answers, emphasize the importance of the computations rather than the correctness of the final answer. Encourage students to explain their solution techniques to gain insight into the nature of any erroneous procedures.

Example 4: A group of college freshmen was given the following problem on a class quiz:

> A McDonald's store has 2000 pounds of french fries delivered. If the maximum capacity of the delivery truck is 120 pounds, what is the least number of truck trips necessary to deliver all the french fries?

Many students in the class divided and obtained the correct solution, seventeen trips (fig. 17.6).

$$\begin{array}{r} 16\,\frac{80}{120} = 17 \text{ trips} \\ 120\overline{)2000} \\ \underline{120} \\ 800 \\ \underline{720} \\ 80 \end{array}$$

Fig. 17.6. Computations for verbal problem in Example 4

When the quiz papers were returned, one student was selected to explain this solution to the class. He proceeded to describe the division process (which was correct) and then concluded, "Since eighty is more than half of one hundred twenty, I round the sixteen up to seventeen," thus revealing that he didn't understand that *any* partial load of french fries, no matter how small, would need another trip. The importance of verbalizing solution techniques as part of an evaluation cannot be overemphasized!

Encouraging students to think aloud as they solve a problem can reveal the thinking that accompanies a solution. Researchers (Alexander 1978; Ashlock 1976; Callahan 1973; Capps 1976; Easley 1977; Inskeep 1978; Lankford 1972; and Underwood 1976) have demonstrated the think-aloud technique as an important diagnostic tool.

Suggestion 4

Listen to students' explanations of the mathematical processes in their solutions. Listening is very helpful, since correct computations are sometimes performed for the wrong reasons.

Example 5: Students in a ninth-grade algebra class were given a diagnostic test that included the multiplication problem 4/5 × 2/3. One solution is shown in figure 17.7.

$$\frac{4}{5} \times \frac{2}{3} = (15)\frac{4}{5} + (15)\frac{2}{3} = 12 \times 10 = 120$$

Fig. 17.7. Some ninth-grade algebra students' computations for 4/5 × 2/3

Probing conducted by the researcher during think-aloud sessions revealed that some students were relating this problem to fractional equations instead of the multiplication of fractions and were applying a technique originally taught to solve fractional equations. Other students demonstrated during the think-aloud sessions that they were attempting to give equivalent fractions a common denominator as an application of the addition-of-fractions algorithm in this problem; they then discovered, "It looks like I'm leaving something out in the whole problem . . . from two fractions how can I get a single answer?" (DeVincenzo 1980, p. 125). It may appear that only the numerators are involved in the faulty algorithm illustrated in figure 17.7, and therefore the algorithm for the addition of fractions is not closely related to the faulty algorithm under discussion. However, it has been observed (DeVincenzo 1980) that students often use the technique of multiplying by a common denominator to arrive at "new" numerators of the equivalent fractions; they then use the common denominator to serve as the denominator of each fraction, thus computing numerators and denominators separately.

Suggestion 5

Use the think-aloud technique in the mathematics classroom to gain insight into the cause of specific errors. Such insight can suggest appropriate remedial activities, since the same faulty algorithm can represent different thinking and thus demand different remediation. Use information from the think-aloud technique to help (*a*) evaluate the sequencing of topics taught, since many errors indicate a confusion between solutions to specific problem types, and (*b*) evaluate the presentation and understanding of specific standard algorithms, since many errors are closely related to solution techniques taught for different problems.

Examples 4 and 5 show that observing students think aloud when they are solving a problem is an excellent way not only to evaluate teaching techniques and student learning but also to suggest remedial activities. Videotaping is another useful diagnostic tool in conjunction with the think-aloud technique: (*a*) it gives teachers an opportunity to examine and evaluate their own participation in the diagnostic technique (their attitude toward students, for instance, and the effectiveness of their questioning and probing) and (*b*) teachers can obtain information from the sequencing of steps during a solution. It should be noted that an analysis of written computations previously completed but not verbally explained and not videotaped cannot reveal the sequencing of steps.

Examples 1 through 5 demonstrate how written computations can be misinterpreted by teachers and suggest that conventional testing does not accurately measure student learning in mathematics. These techniques to improve teachers' diagnostic procedures are presented to help improve the teaching and learning of mathematics in the 1980s.

REFERENCES

Alexander, David. "A Clinical Diagnostic Investigation of Fundamental Reasons for Conceptual Difficulties with Algebra." (Doctoral dissertation, State University of New York at Buffalo, 1977.) *Dissertation Abstracts International* 39 (1978):764A. (University Microfilms No. 78-13,995)

Ashlock, Robert. *Error Patterns in Computation: A Semi-Programmed Approach.* Cincinnati, Ohio: Charles E. Merrill Publishing Co., 1976.

Callahan, Leroy G. "Clinical Evaluation and the Classroom Teacher." Paper presented at the AERA meeting, 25 February–1 March 1973, in New Orleans. (ERIC Document Reproduction Service No. ED 076 640)

Capps, Lelon. "Thoughts on Coordinating a Research Effort in Remediation in Mathematics." In *Proceedings of the Third National Conference on Remedial Mathematics,* edited by James Heddens and Frank Aquila. Kent, Ohio: Kent State University, 1976.

Cox, Linda. *Analysis, Classification and Frequency of Systematic Error Computational Patterns in the Addition, Subtraction, Multiplication and Division Vertical Algorithms for Grades 2–6 and Special Education Classes.* Kansas City, Kans.: Kansas University, 1974. (ERIC Document Reproduction Service No. ED 092 407)

DeVincenzo, Mary Ann. "An Investigation of the Relation between Elementary Algebra Students' Errors in Arithmetic and Algebra in Selected Types of Problems." (Doctoral dissertation, New York University, 1980.) *Dissertation Abstracts International* 41A (1980). (University Microfilms No. 80-17,494)

Easley, J. R. *On Clinical Studies in Mathematics Education.* Champaign, Ill.: University of Illinois at Urbana-Champaign, 1977.

Epstein, Marion G. "Testing in Mathematics: Why? What? How?" *Arithmetic Teacher* 15 (April 1968): 311–19.

Graeber, Anna O., and Lisa Wallace. *Identification of Systematic Errors: Final Report.* Philadelphia: Research for Better Schools, 1977. (ERIC Document Reproduction Service No. ED 139 662)

Inskeep, James E. "Diagnosing Computational Difficulty in the Classroom." In *Developing Computational Skills,* 1978 Yearbook of the National Council of Teachers of Mathematics, edited by Marilyn N. Suydam. Reston, Va.: The Council, 1978.

Lankford, Francis. *Some Computational Strategies of Seventh Grade Pupils, Final Report.* Charlottesville, Va.: University of Virginia, 1972. (ERIC Document Reproduction Service No. ED 069 496)

MacKay, I. D. *A Comparison of Students' Achievement in Arithmetic with Their Algorithmic Confidence.* Vancouver, B.C.: Mathematics Education Diagnostic and Instructional Centre, British Columbia University, 1975. (ERIC Document Reproduction Service No. ED 128 228)

Pincus, Morris, Margaret Coonan, Harold Glasser, Lillian Levy, Frances Morgenstern, and Herbert Shapiro. "If You Don't Know How Children Think, How Can You Help Them?" *Arithmetic Teacher* 22 (November 1975): 580–85.

Underwood, James. "An Exploratory Study of the Problem Solving Procedures Used by Selected College Freshmen on Certain Basic Consumer Mathematics Problems." (Doctoral dissertation, University of Tennessee, 1976.) *Dissertation Abstracts International* 37 (1976): 5665A. (University Microfilms No. 77-3694)

West, Tommie A. "Diagnosing Pupil Errors: Looking for Patterns." *Arithmetic Teacher* 18 (November 1971): 467–69.

18

Factors That May Influence Performance on Standardized Tests

Margaret McDonald

DURING the 1980–81 school year, teachers in the Nelson School in Kansas City participated in a special project that had as its primary goal the improvement of student scores in mathematics on the standardized achievement test given each year in the system. At the beginning of the school year, all the teachers in the school (grades 1–6), the principal, and a consultant met to develop a plan of action.

They decided that the item analysis printouts of test results from the previous year could be used in an attempt to discern any general deficiencies in student achievement. These results could be used to plan instructional improvements. The test manual provided a list of objectives, with each test item keyed to an objective. Thus, although the test was a norm-referenced test, a careful examination of test results item by item would actually provide a kind of criterion-referenced feedback.

Of course, the first step was to verify that the test objectives were consistent with the school system and school objectives. After some comparison and discussion, it was agreed that in general the test objectives were compatible with local objectives. The next step involved a careful examination of the item analysis results. As a working guide, it was decided that any item on which fewer than 60 percent of the students responded correctly merited closer examination relative to the objective it purported to measure. Furthermore, poor performance on several items keyed to the same objective was assumed to give ample evidence of a concept or skill that should receive particular attention in the instructional program. In addition, careful note was made of any deficiencies that seemed to crosscut two or more grades.

The extended plan was to use these identified deficiencies as the basis for a series of workshops with the teachers. These workshops were to assist teachers in developing instructional strategies and materials that would address the particular needs identified in the test analysis. It was agreed that such a plan should lead to improved achievement on test scores. The consultant was asked to assist in the analysis and to lead in designing workshops around the topics that concerned the teachers.

The initial task of identifying the test items with which students had been least successful was undertaken. The available item analysis printouts were used to flag each item that met the criterion of failure amounting to 40 percent or greater. The actual item was examined in reference to the objective it purported to measure. Then an attempt was made to identify the specific concept or skill that contributed to the students' inability to respond correctly to the item.

Some interesting general patterns did begin to emerge early in the analysis. As expected, there was evidence of areas that were causing difficulty for a large number of students. These areas became the basis for instructional planning for remediation. Among those evident early in the analysis were fractions (including decimal fractions in the upper grades) and word problems at all grade levels. These topics became the focus for several workshops.

One rather specific difficulty identified in the analysis was the general failure of students to use correctly a fraction associated with a set model. For example, when students were asked to identify a representation of one-fourth of twenty objects pictured, their answers were almost randomly distributed over the available choices. This was taken as fairly good evidence of a general lack of understanding of the model. As a part of the workshop activities, materials and instructional strategies were suggested to help recognize the set model for fractions. This is only one of several rather specific concepts that were identified for special attention.

In addition to those areas of difficulty that could be identified and improved, it became apparent almost immediately that there were other factors that influenced performance on some items and that these factors seemed unrelated to the concept or skill being tested. One of the first apparent deterrents to performance on an item at all grade levels was the inclusion of a negative stem in the item. For example, any item that asked the student to identify the answer that did *not* satisfy the stated conditions uniformly elicited a low success rate on the part of students. Although the item might purport to measure knowledge of a fraction model or meaning, it really seemed to test the ability of the student to handle the negative stem.

To verify this hypothesis, some practice items were designed that tested the same concept in a variety of multiple-choice as well as single-response forms. Included in the items were a few with negative stems. As expected, students performed poorly on those items that used a negative form—for example, the item that appears in figure 18.1. Many students selected response (*a*), which was the first correct representation of one-half. On similar items with positive stems they responded correctly.

Which of these is not a picture of $\frac{1}{2}$?

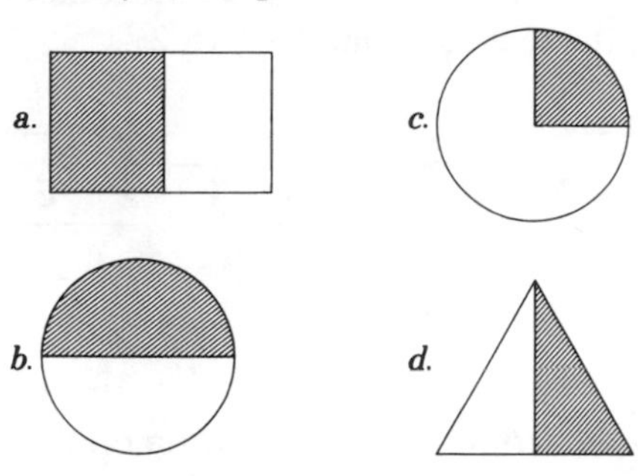

Fig. 18.1

Since textbooks seldom include questions that require the identification of items that do not belong to a particular response, students do not have experience in selecting wrong answers. Furthermore, teachers also pointed out that in general it was not considered good teaching strategy to show incorrect answers. It was agreed, however, that if students were to be expected to recognize nonexamples, experience with selecting incorrect answers could contribute to a student's grasp of the concept. Supplementary lessons, therefore, were designed at all levels to give students experience with the troublesome negative.

There were other seemingly trivial discoveries in the analysis that turned out to be not so trivial as far as student responses were concerned, as evidenced by the frequency with which certain items were missed. The greatest distractions seemed to be in the discrepancies between the mathematical language of the particular text series being used and the language of the test. Such differences as *numeral* in the text and *name for a number* on the test seemed to be a source of confusion for some students. Often, it was not a matter of correctness of language but rather the students' familiarity with the particular word used.

Other variations in language and models that seemed to cause students difficulty are these:

1. *Equivalent* was used on the test, as in equivalent sets or equivalent forms of a numeral (that is, using expanded notation). The only use of *equivalent* in the students' text series was in relation to equivalent fractions. Here again, both are correct, but the students' experience with the form of the word used on the test was extremely limited.
2. *Operator* was used on the test to refer to an operation as indicated by the symbol for the operation. This was not a part of the vocabulary of the text.
3. In subtraction, *rename* was used on the test, but the text used *regroup* exclusively.

4. The form of the number line model for subtraction differed. The test used two arrows; the text used one (fig. 18.2).

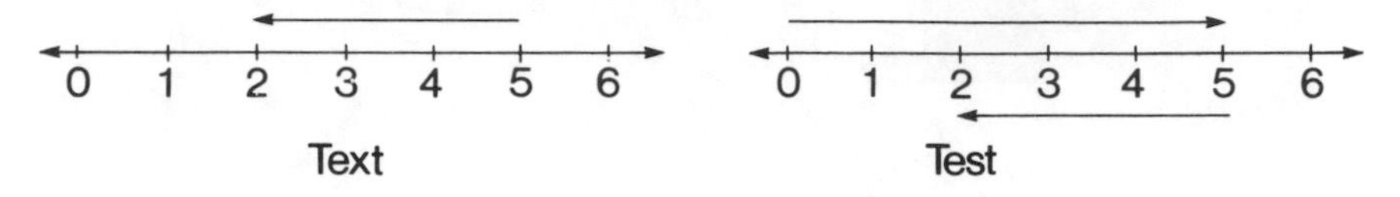

Fig. 18.2

5. Fractions were to be written in *simplest* form on the test, but the text asked for *lowest* terms.

Although such variations in language and form may not seem significant at first glance (and this is only a sample of observed differences), it was felt that the lack of success on items including them was convincing enough to warrant giving attention to them in the instructional program. It was agreed, therefore, that language, including variations in terminology as well as models for concepts, would have to receive greater attention at all grade levels.

The results of the teachers' efforts were rewarding, both in the emphasis on concepts and skills with which students had evidenced difficulty as well as attention to variations in vocabulary and models. End-of-the-year norm-referenced test results showed substantial improvement at every grade level over the previous year's results.

At no time was it felt that the project involved "teaching to the test." In fact, every decision about the expenditure of time, both of teachers and students, was predicated on whether the instructional activity contributed to the realization of a goal that was acceptable to the entire staff. Even if the test scores had not shown appreciable gain, it was agreed by all who participated that instruction in mathematics improved. Students and teachers evidenced satisfaction with the changes in their program.

The project was not designed as a research study, and the observations were entirely subjective. However, all who participated developed an increased sensitivity to factors that influence success on a standardized test and an appreciation of the need to (a) use as many models as possible for a particular concept and (b) pay particular attention to variations in vocabulary. It was agreed that the improved test scores were a concomitant of improved teaching.

There also remained an acute awareness of the potential influence that subtle item characteristics on a nationally developed test might have on test scores. Teachers who participated in the analysis have a heightened sensitivity to these characteristics and recommend a similar critical analysis to others.

19

Assessing and Improving a School's Mathematics Program, K–8

Nancy C. Whitman

THE State of Hawaii Department of Education developed and piloted a school-level mathematics assessment and improvement model to help schools evaluate their mathematics program. Several problems relating to evaluation had to be resolved in the development process. These were mainly problems of how to incorporate appropriate evaluation procedures into the schools' assessment and improvement activities—problems, for instance, of how to—

1. make it imperative that the program's goals determine the type of instrument needed to assess program effectiveness, student learning, and quality of materials;

2. ensure that the evaluation of mathematics learning include the full range of the program's goals, including the reading, content, and processes of mathematics;

3. make the use of nontest data an integral part of the assessment and improvement activities of school mathematics programs;

4. ensure that the evaluation of materials be an integral part of program planning;

5. involve classroom teachers and school administrators in the assessment and improvement of their school-wide mathematics program.

These problems correspond closely to those of effectively implementing Recommendation 5 of the *Agenda for Action.* What follows is a discussion of this correspondence between the instruments that were developed in Hawaii to help schools assess and improve their mathematics programs and the implementation of various recommended actions for Recommendation 5. The instruments, called schedules, will be explained and their corresponding recommended actions noted. Several will appear here in whole or in part.

Briefly, they cover the following topics:

Schedule A. School Profile
Schedule B. Grade-Level Assessment
Schedule C. Summary Sheet
Schedule D. Summary of Needs
Schedule E. Planning for Improvement
Schedule F. Mathematics Programs of Assistance
Schedule G. Suggestions for Implementation

Implementing Recommendation 5 Through Instruments Formulated for Assessing and Improving Hawaii's Schools

The instruments are intended to be used repeatedly over a period of years. Initially, however, they should be used so that a school can complete its improvement plan, using Schedule E, and begin implementing the plan, using Schedule G, within one year.

Recommended Action 5.1: *The evaluation of mathematics learning should include the full range of the program's goals, including skills, problem solving, and problem-solving processes.*

Schedule B, the grade level assessment, was used to assess the grade-level mathematics program relative to content covered, procedures experienced, and test results. A one- to four-page form is available for each grade K–6, and a four-page form for grades 7–8. Schedule B was developed to incorporate the content objectives and the procedural goals from Hawaii's *Mathematics Program Guide* (1979). Furthermore, the schedule provides for the analysis of the mathematics content and processes concurrently. Schedule B for grade 3 is shown in figure 19.1. The assessment for the other grade levels uses the same general format.

Schedule C, a summary sheet, emphasizes the importance of both content and process goals by summarizing in two major parts the information obtained by grade level in Schedule B. One part summarizes the content covered in grades K–8 and the test results that are available; the other part summarizes the learning experiences provided for students in the different content areas. Such experiences include problem solving, communication and computation, and using manipulatives and other forms of mathematical representation. Portions of Schedule C appear in figure 19.2.

Recommended Action 5.3: *Teachers should become knowledgeable about, and proficient in, the use of a wide variety of evaluative techniques.*

In Hawaii, the successful use of the assessment and improvement instruments required the involvement of classroom teachers and school administrators. Both groups needed training in a variety of evaluation techniques. The fact that assessment means more than obtaining standardized test results and that nontest results provide important information in assessing a school's mathematics program had to be emphasized.

Schedule A provided a profile of the school. It was developed to help teachers and administrators realize the role and importance of nontest data. The information obtained in Schedule A included data on the following:

1. The school
   - *a*) Student enrollment
   - *b*) Number of regular classroom teachers at each grade level
   - *c*) Number of other professional personnel
2. The student body
   - *a*) Percent of student body that is part-Hawaiian, Korean, Chinese, and so on
   - *b*) Percent of student body for which English is a second language
   - *c*) Percent of student turnover
3. The professional mathematics background of the teachers
4. The school budget
5. School procedures relative to mathematics
   - *a*) How the annual mathematics materials budget is determined
   - *b*) How texts and other instructional materials are selected
   - *c*) Which groups have discussed and studied the school mathematics program in relation to the various statewide guidelines
6. Testing and student evaluation
   - *a*) Whether there is a master list of school test results
   - *b*) What tests have been given by the school, and how they have been used in placing students; assessing mathematics programs and student needs
   - *c*) Whether test results of the past few years indicate an improvement or a decline in the students' knowledge and skills
7. Needs of special students
   - *a*) Methods recommended by the school for meeting the needs of such students as poor readers, nonreaders, recent immigrants, and mainstreamed handicapped students
   - *b*) How better mathematics students are being challenged

The information obtained in Schedule A served as a basis for analyzing data on subsequent schedules and for formulating the school's improvement

Number of days per week mathematics is taught: _______

Length of period: _______

Reading level of materials for students: _______

Is reading level appropriate? _______

	LEARNER OBJECTIVE COVERED	NO. CLASS PERIODS OF INSTRUCTION (YEAR)	COVERED IN TEXTBOOK	PROCEDURES: A. PROBLEM SOLVING	B. COMMUNICATION SKILLS	C. RELATES TO STU-DENT'S ENVIRONMENT	D. MANIPULATIVES & REPRESENTATIONS	E. INDEPENDENT INVESTIGATIONS	F. COMPUTATIONAL SKILLS	TEST RESULTS
NUMBERS AND OPERATIONS:										
• Orders numbers to 1000 (1)*										
• Counts by 100's, 10's, 5's, 4's, 3's, 2's, and 1's (4)										
• Writes numerals and names place value										
• Reads and writes roman numerals (2)										
• Uses ordinal and cardinal numbers										
• †Number properties (maintains and extends) (5)										
• Adds and subtracts 3-digit numbers										
• Adds three addends (3)										
• Adds and subtracts without paper and pencil										
• Uses number line for multiplication and division										
• Uses relationship of multiplication and division										
• Finds product of three factors (6)										
• Relates multiplication to addition and division to subtraction										
• Uses facts through products of 81										
• Multiplies with factor less than 100										
• Divides with one-digit divisor										
• Identifies two-thirds, three-fourths										
• Reads and writes fractions (8)										
• Uses terms *numerator* and *denominator*										
• Adds and subtracts fractions using pictures										

*Numbers in parentheses correspond to those in Schedule C: Summary Sheet: Content Coverage Grades K–6.

†An explicit objective of grades 2 and 5 in the *Mathematics Program Guide* and an implicit objective of grades 3 and 4.

Number of days per week mathematics is taught: ________

Length of period: ________

Reading level of materials for students: ________

Is reading level appropriate? ________

	LEARNER OBJECTIVE COVERED	NO. CLASS PERIODS OF INSTRUCTION (YEAR)	COVERED IN TEXTBOOK	PROCEDURES: A. PROBLEM SOLVING	B. COMMUNICATION SKILLS	C. RELATES TO STUDENT'S ENVIRONMENT	D. MANIPULATIVES & REPRESENTATIONS	E. INDEPENDENT INVESTIGATIONS	F. COMPUTATIONAL SKILLS	TEST RESULTS
GEOMETRY:										
• Names and draws geometric figures (18)										
• Investigates geometric properties										
• Investigates line symmetry (19)										
• Draws congruent figures										
• Understands geometry terminology (20)										
• Locates and forms angles (21)										
MEASUREMENT:										
• Tells and writes time in minutes (26)										
• Describes a year in months, days										
• Reads and writes money expressions										
• Makes change (27)										
• Counts and records money amounts										
• Describes boiling and freezing points of water on thermometer (28)										
• Investigates gram and kilogram (29)										
• Finds perimeter (30)										
• Explores area (33)										
• Measures volume in centiliters and milliliters										
• Forms figures and tells volume (34)										
• Investigates relationship of capacity and volume units										
• Draws graphs (35)										

Fig. 19.1. Schedule B: Grade 3

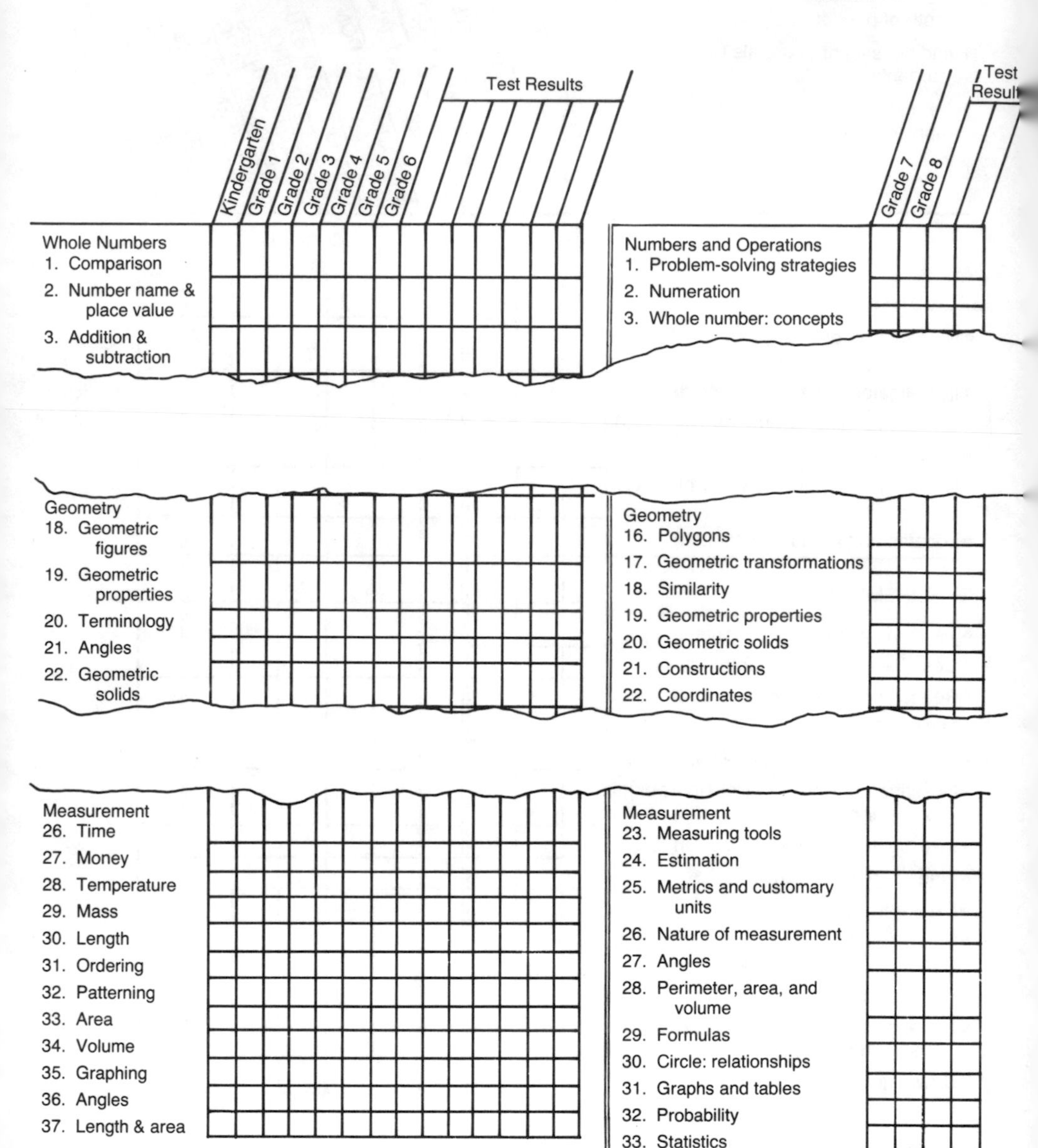

SCHEDULE C
Summary Sheet: Content Coverage
Grades K–6

	Kindergarten	Grade 1	Grade 2	Grade 3	Grade 4	Grade 5	Grade 6	Test Results						
Whole Numbers														
1. Comparison														
2. Number name & place value														
3. Addition & subtraction														
Geometry														
18. Geometric figures														
19. Geometric properties														
20. Terminology														
21. Angles														
22. Geometric solids														
Measurement														
26. Time														
27. Money														
28. Temperature														
29. Mass														
30. Length														
31. Ordering														
32. Patterning														
33. Area														
34. Volume														
35. Graphing														
36. Angles														
37. Length & area														

SCHEDULE C
Summary Sheet: Content Coverage
Grades 7–8

	Grade 7	Grade 8	Test Results	
Numbers and Operations				
1. Problem-solving strategies				
2. Numeration				
3. Whole number: concepts				
Geometry				
16. Polygons				
17. Geometric transformations				
18. Similarity				
19. Geometric properties				
20. Geometric solids				
21. Constructions				
22. Coordinates				
Measurement				
23. Measuring tools				
24. Estimation				
25. Metrics and customary units				
26. Nature of measurement				
27. Angles				
28. Perimeter, area, and volume				
29. Formulas				
30. Circle: relationships				
31. Graphs and tables				
32. Probability				
33. Statistics				

SCHEDULE C

Summary Sheet: Procedures Experienced

Grades K–8

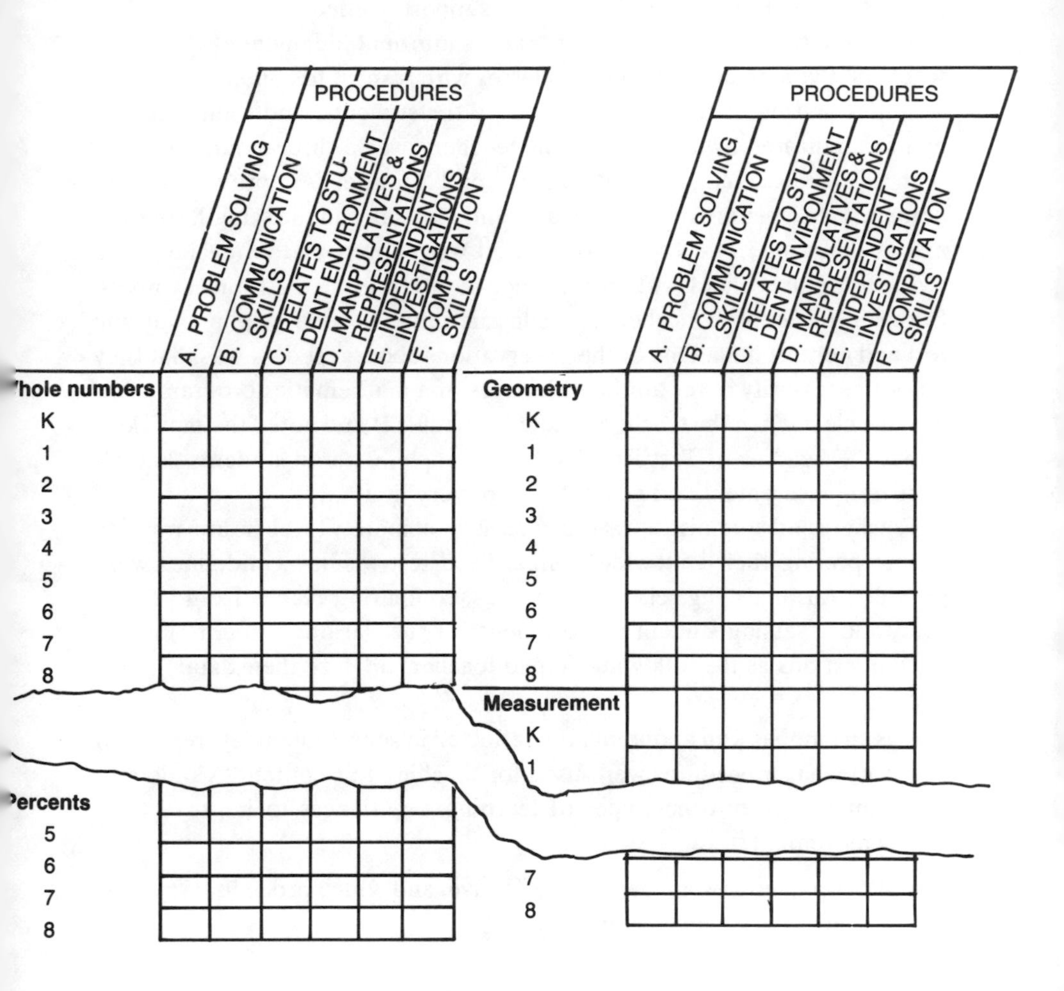

	PROCEDURES					
	A. PROBLEM SOLVING	B. COMMUNICATION SKILLS	C. RELATES TO STUDENT ENVIRONMENT	D. MANIPULATIVES & REPRESENTATIONS	E. INDEPENDENT INVESTIGATIONS	F. COMPUTATION SKILLS
Whole numbers						
K						
1						
2						
3						
4						
5						
6						
7						
8						
Percents						
5						
6						
7						
8						

	PROCEDURES					
	A. PROBLEM SOLVING	B. COMMUNICATION SKILLS	C. RELATES TO STUDENT ENVIRONMENT	D. MANIPULATIVES & REPRESENTATIONS	E. INDEPENDENT INVESTIGATIONS	F. COMPUTATION SKILLS
Geometry						
K						
1						
2						
3						
4						
5						
6						
7						
8						
Measurement						
K						
1						
7						
8						

Fig. 19.2

plan. For example, the data on the socioeconomic and ethnic background of students needed to be considered in comparing local school tes results to the national norm. The background information for mathematics teachers provided clues to the type of in-service support needed.

Schedule B (fig. 19.1) recognizes that the informal judgment of classroom teachers is vital in evaluating their classes with respect to content coverage, activities and procedures experienced by the students, and their achievement as indicated by test results. The teachers used a three-point system—such as E for excellent, S for satisfactory, and N for needs improvement—to help provide a picture of the emphases and omissions in the school's mathematics programs. The system to be used was chosen by the teachers as a group to ensure school-wide consistency in interpreting the rating symbols. In this instance, both teachers and administrators needed help in analyzing test and nontest data. One of the observations about test data was that they do not necessarily cover all the objectives of a mathematics program. This became clear when the teachers studied Schedule B and noted the omissions under the heading of Test Results. For example, data on students' knowledge of measurement and geometry were very limited.

Teachers and administrators received approximately three hours' training in interpreting their statewide standardized test results. Guidelines were provided for analyzing school summaries, comparing results from year to year, and assessing student achievement for the various content clusters. Such questions as the following helped teachers analyze their data:

- Is an emphasis on a content area reflected in achievement test results? If not, what procedures were used for teaching this content? Could students be given other types of learning experiences to improve their achievement?
- What procedures appear to be effective, and which procedures may be ineffective for a given content area?

Schedule D, a summary of needs (fig. 19.3), was developed to further help schools analyze their data and plan for improvement. This schedule organizes and summarizes the analysis of the school's mathematics program. It helps the school to use test and nontest data to study the school's program relative to content, procedures, and student evaluation and to identify and rank aspects of the mathematics program that need intervention, such as in-service training. The analysis is based on the school-wide profile reflected primarily in Schedules A and C.

SCHEDULE D
SUMMARY OF NEEDS

Schedule D should reflect those aspects of the school's mathematics program (e.g., curriculum, staff, instructional materials, and so on) that require intervention.

Use the information on Schedules A, B, and C, the policies of the school, district, or state that are related to mathematics education, your own experiences, and relevant information from other sources to complete this schedule.

I. *Program Analysis*

To complete this section, refer to Schedule C. If information for a particular grade level is required, refer to Schedule B.

A. Content

1. Identify content that needs to be deleted.
2. Identify new content that needs to be incorporated into the mathematics program.

B. Procedures

Identify procedures that need to be incorporated into or emphasized in the mathematics program.

C. Student Evaluation

Identify content areas in which test results do not reflect the emphasis given to them in the classroom.

II. *Analysis of Other Areas*

List additional aspects of the mathematics program that require intervention.

A. Student groups whose needs are not being met in a satisfactory manner.
B. Administrative procedures (e.g., budget, school liaison).
C. Curriculum/program concerns (e.g., instructional materials).
D. Staff (e.g., in-service needs).
E. Other.

Fig. 19.3

Recommended Action 5.4: *The evaluation of mathematics programs should be based on the program's goals, using evaluation strategies consistent with these goals.*

Schools that use these schedules for evaluating their mathematics programs are essentially basing their assessment on the program's goals. The objectives in Schedule B are basically the objectives stated in Hawaii's *Mathematics Program Guide.* The procedures enumerated in Schedules B and C are also listed in the *Mathematics Program Guide.* The questions in Schedule B dealing with the reading level of mathematics materials reflect the statewide mandate that the reading of subject matter be taught in the content areas.

Although the results of the statewide testing using standardized tests are available, only those items that are relevant to the objectives and goals of Hawaii's mathematics program are recorded in Schedule B. Furthermore, the scarcity or lack of items that measure the objectives in Schedule B is also noted. Because of the limitations of the standardized tests being used, schools have sought more appropriate tests. A few schools have created their own.

These two instruments (Schedules B and C) have helped teachers and administrators focus their attention on those objectives and experiences considered important in the *Mathematics Program Guide* and in other statewide guidelines.

Schedule D, a summary of needs, focuses on the procedures that need to be emphasized and the content that needs to be incorporated or deleted from the program. Calling these needs to the school's attention keeps test scores from being the sole index of success. Also, the use of informed teacher judgments in evaluating student learning in Schedule B, and of nontest data found in Schedule A, make test scores just one of many factors to be considered in evaluating the school's mathematics program.

Recommended Action 5.5: *The evaluation of materials for mathematics teaching should be an essential aspect of program planning.*

Schools that identify teaching materials as a need of their mathematics program and plan to address that need will find Schedule F useful. Schedule F helps schools in their program planning by bringing to their attention recent and widely used textbooks and alternatives to textbooks. Schedule F also identifies features that are common to the textbooks, especially those that touch on the goals and objectives of Hawaii's mathematics program. It cites learning experiences that may be new to mathematics teachers or different from those presently in use. These ideas on teaching strategies and student experiences provide additional background for the analysis of school's mathematics programs. Its brief descriptions and analyses of mathematics materials give schools further background to add to their summary of needs.

Since the descriptions are intended as a beginning from which materials can be selected for a thorough study, some of these programs may be identified as part of the school's improvement plan (Schedule E). This schedule contains a format for developing a three-year improvement plan for the school. Each statement of need is analyzed relative to such considerations as the resources it requires and the constraints it imposes. A time line is established for addressing the needs and implementing the improvements.

Mathematics program planning should be a school-wide activity. Hence, the selection of textbooks should also be a school-wide activity involving both teachers and administrators. Further, since program assessment is based on a program's goals and objectives, the selection of textbooks for improving a program is also based on the program's goals and objectives. Schedule F can facilitate this process, since it analyzes programs in relation to goals. In addition, Schedule B may be used to analyzc instructional materials for coverage of the *Mathematics Program Guide* objectives. This analysis is useful when reviewing new materials or programs in mathematics for school adoption.

Changes Observed in the Schools

Changes relative to Recommendation 5 have been observed in the schools that piloted these instruments:

1. Some of the schools were able to pinpoint both content *and* process deficiencies in their students. They were able to identify in their improvement plans (Schedule E) procedural goals that needed attention. Statements such as "need to attend to problem solving," "need to incorporate manipulatives in instruction," "need to provide provisions for independent investigations by students," and "need to relate mathematics to the students' environment" were seen in the improvement plans.

2. Some of the schools were able to diagnose particular content and process needs in their students, and they were able to elicit and obtain help with these needs. A few of the schools implemented plans to evaluate their curriculum changes. Several schools identified the need for in-service training in teaching and evaluating process skills such as problem solving.

3. Many teachers became aware of their school's total mathematics program rather than just that portion for which they were directly responsible. Long-range plans based on school-wide assessments using Schedules A through D were developed and implemented.

4. Both teachers and administrators became more program oriented and less textbook oriented. They became more knowledgeable about mathematics programs and resources and about the means for obtaining assistance with these programs. The selection of textbook materials became a school-wide involvement of teachers and administrators.

Other notable changes seen in some of the schools included (*a*) increased communication among teachers and between teachers and administrators, (*b*) increased knowledge about the state of Hawaii's *Mathematics Program*

Guide, and (*c*) increased ability to relate the school's mathematics program to that of the *Mathematics Program Guide.*

A wider range of measures than conventional testing will give schools information for sound decisions as they continue to evaluate the success of their mathematics programs and the achievement of their students. They will be more accountable to their public, and they will know how well they are doing in the broad spectrum of mathematics learning.

BIBLIOGRAPHY

Hawaii Department of Education, General Education Branch. *Mathematics Program Guide.* Honolulu: The Department, 1979.

———. *State Reading Improvement Framework.* Honolulu: The Department, 1979.

National Council of Teachers of Mathematics. *An Agenda for Action: Recommendations for School Mathematics of the 1980s.* Reston, Va.: The Council, 1980.

Whitman, Nancy C. *Final Report of the Improvement of School Programs in Mathematics through a Comprehensive Foundation Program Assessment and Improvement System (FPAIS) Approach Project (1977-1980).* Honolulu: College of Education, University of Hawaii, 1980.

Whitman, Nancy C., Evelyn Horiuchi, David Tanner, John Marks, Morris Lai, and Ruth Wong. *Mathematics: Program Assessment and Planning Manual.* Honolulu: College of Education, University of Hawaii, 1980.

20

An Alternative to Conventional Methods of Evaluation

Hilde Howden

RECOMMENDATION 5 from the *Agenda for Action* strikes at the deeply rooted tradition of paper-and-pencil evaluation in mathematics. Several factors contribute to this tradition. We have been conditioned by our own school experiences, by the emphasis on standardized testing, by advances in mechanical grading techniques, by public stress on computation, and by the demand for documented accountability. Most insidious of all the contributing factors is the popular misconception of mathematics itself. Mathematics is the only academic school subject that comprises both concepts and skills; yet it is generally taught, and consequently tested, as if it were merely a body of technical procedures.

The Albuquerque public schools are attempting to change both the perception of what mathematics is and how a student's knowledge of mathematics can be evaluated more meaningfully. Four years ago, the District Board of Education mandated key competencies in mathematics. The teacher committee that had developed the district's Key Competencies in Mathematics Project accepted the mandate as a challenge to effect some major changes in the mathematics program. They decided to—

- initiate an awareness of the dual nature of mathematics;
- apply the results of recent learning theory research;
- incorporate teacher observation as an evaluation tool;
- place more responsibility for learning on students by making them and their parents aware of identified expectations;
- monitor more effectively students' progress in mathematics from grade to grade to provide for individual student needs.

Although the provisions for achieving all these objectives are interwoven throughout the project, the primary emphasis of this report is on observation as an essential evaluation tool.

Student "Math Paks" form the core of the project. In grades 1–5, the Math Pak for each grade level consists of a set of six colored cards on a ring (fig. 20.1). Each child has a set, and each card in the set lists the grade-level competencies for one of the major strands of the mathematics curriculum. Each card also has extra spaces in which to write additional objectives for individualizing instruction. Deficiencies from previous grade levels, competencies in greater depth, and special projects are examples of the kinds of objectives that may be listed in these spaces. As mastery of any one of these objectives is observed, it is recorded in the student's Math Pak (fig. 20.2).

Fig. 20.1. The Math Pak is a constant reminder to teachers, students, and parents of learning expectations.

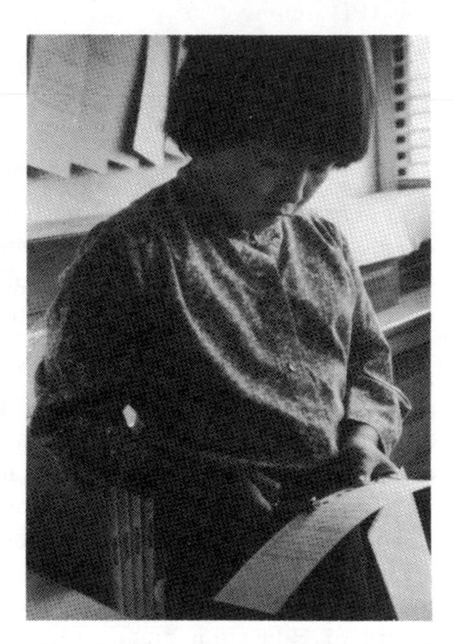

Fig. 20.2. By punching their own Math Paks, students become more involved in their learning process. The Math Paks in the background are hanging on one of the racks that Principal Ongelo Pierson made for all the classrooms at Mountain View School.

In the middle school (grades 6–8), the Math Pak takes the form of a Math Pak Sheet. Students are given a copy and are encouraged to use it for self-evaluation.

The prominently displayed Math Paks in elementary classrooms act as a constant reminder to students of their individual learning expectations.

Contrary to early fears, students rarely compare Math Paks but instead strive to achieve competencies so as to get their Paks "punched" and have additional learning goals added. Displaying the Math Paks also acts as a constant reminder to teachers to watch and listen as students interact in their mathematics activities.

To assist teachers in such observed evaluation, the Teacher's Resource Book, also developed by teachers in the district, includes many suggestions for activities that are correlated to the competencies. Learning activities such as calculator investigations, instructional games, and modeling with manipulatives are organized in an easy-to-use matrix format. The Diagnostic/Assessment section of the Teacher's Resource Book gives four suggestions for each competency. They include one activity for peer interaction (PI) among two or three students, a student-teacher (ST) activity intended for one-on-one evaluation, an independent work (IW) activity that students perform individually, and a group communication (GC) activity designed to encourage students to talk about mathematics. For example, the four suggested diagnostic/assessment activities for the competency

The student will subtract two-digit numbers with regrouping

follow:

- PI Students work in pairs. One student writes a number on a sheet of paper or on the chalkboard. Under it, the second student writes a smaller number. The first student writes their difference; the second student checks the answer by adding the subtrahend and difference. Roles are reversed after each check is completed.
- ST Ask the student to demonstrate with bean sticks, numeration blocks, or Sigma Grids the solution to a problem such as 342 – 236.
- IW The student does a worksheet or a textbook exercise set and uses manipulatives such as bean sticks, numeration blocks, or Sigma Grids to check the work or as an aid when necessary.
- GC Students work together in groups of about six to find the missing digits in previously prepared problems such as those shown below. As they work together, students should discuss the rationale for assigning digits to the blanks.

```
  5□75        74□□
- 38□4      -  □59
  □86□        □087
```

Two primary purposes underlie the inclusion of such suggestions for each competency. They encourage the use of more learning activities that require students to talk about mathematics, thus promoting greater understanding of the concepts of mathematics. They also provide for an atmosphere in which the teacher can truly observe the students' understanding and learning progress (fig. 20.3).

In addition, the Math Paks and the Diagnostic/Assessment suggestions provide a communication link between a student's regular classroom teacher and substitutes, aides, teachers of special programs, and parents (fig. 20.4). This link makes them all aware of specific needs, learning activities to meet those needs, and ways to evaluate progress.

Fig. 20.3. Fourth-grade Teacher Julia Howington observes as students use bean sticks to model the addition of three-digit numbers.

Fig. 20.4. Title I Teacher Margaret Clark recognizes the mastery of computational skills while observing students engaged in an instructional game.

The communication link also extends to the other schools in the district, which is especially important in the Albuquerque area, where a mobility rate of up to 80 percent is a way of life. The Math Paks help in grouping students with similar needs, whether they be remedial, maintenance, or extension, and the Diagnostic/Assessment suggestions provide activities to meet those needs. An activity from a previous grade level, for example, can be used for reteaching or reinforcement and is as accessible as an enrichment activity from a succeeding level.

To ensure that mastery is retained, an Individual Record Card monitors student progress from grade to grade. Retained mastery is evaluated at the end of each year by a criterion-referenced level test. In grades 1 and 2, the level tests consist of a nonwritten part, which can be administered to the whole class, and a written part. Observed mastery, retained mastery, or no mastery of each competency is recorded at the end of each school year on the student's Individual Record Card. Information from these cards is used to evaluate the project, to identify learning styles and disabilities, to design remediation, and to improve communication between grade level teachers at a given school and between elementary and middle schools.

After a pilot year and just two years of district-wide use in seventy five elementary and twenty two middle schools, the effectiveness of the project is already evident. It has become a unifying element for mathematics education in the district, a focus for teacher in-service training, and a basis for a summer reinforcement program. But best of all, the project is beginning to change attitudes about the need for understanding mathematics as opposed to simply performing rote manipulations and how this understanding can be evaluated by methods other than conventional testing.

The success of the project has resulted in two major extensions of the program. It has been developed for use in grades 9–12, and the Individual Record Card has been computerized for easier access and transfer of information.

Recommendation 6

MORE MATHEMATICS STUDY MUST BE REQUIRED FOR ALL STUDENTS AND A FLEXIBLE CURRICULUM WITH A GREATER RANGE OF OPTIONS SHOULD BE DESIGNED TO ACCOMMODATE THE DIVERSE NEEDS OF THE STUDENT POPULATION

21

An Enrichment Program for Students of Exceptional Capability

Eugene Jercinovic

THE Albuquerque Public Schools district serves an area of 1 243 square miles with ten high schools, twenty-two middle schools, seventy-five elementary schools, and six alternative schools. The 1980–81 enrollment was 78 950. The large size of the district and the wide variations in student achievement have provided continuous challenges to developing curricula to meet the needs of the entire student population.

The district completed its implementation of the elementary, middle, and four-year high school form of organization in 1977–78. With the introduction of the middle school to the system, the fundamental philosophy and organization of the mathematics curriculum was reexamined, and wide-based enrichment and exploration of mathematics has emerged as the guiding philosophy of the middle school. In 1978–79 the teaching of algebra at the eighth-grade level was discontinued. With the cooperation of teachers throughout the district, a new honors curriculum was written for sixth, seventh, and eighth grades. Although containing a strong prealgebra strand, the curriculum also placed considerable emphasis on many other areas of mathematics, such as probability, inductive and deductive reasoning, sets, trigonometric ratios, and geometry. In most cases this curriculum is effectively meeting the needs of mathematically able students at the middle school level.

During the past two years, however, a need has been recognized for a stronger program for a small group of exceptionally capable students spread thinly throughout the district. These students demonstrate extremely rapid learning rates and high retention. The special mathematical needs of this group are met in part by the regular classroom teachers, but a clear need has existed to provide a strong content-based supplementary program for middle school students. The district's response has been to develop the Middle School Honors Seminar in Mathematics, an itinerant program in which the instructor travels and meets with unusually advanced students once a week in their own schools.

Of paramount importance in carrying out such a program was the development of a suitable identification process. Criteria for eligibility were developed and include the following: high ability as indicated on tests of mathematical aptitude, relatively high performance on tests of mathematical achievement, and educational maturity as indicated by teachers in the school. Initial screening of the middle school population was accomplished by identifying students who scored in the top 2 percent on the numerical ability, abstract reasoning, and space relations subtests of the Differential Aptitude Test (Psychological Corporation, New York). These tests are administered by most sixth- and seventh-grade mathematics teachers as a placement instrument for their own honors programs. Separate testing can be arranged for transfer students, students who were absent during testing, elementary level students, and for retesting those students whose previous DAT scores were very close to qualifying them for the program. Records of students thus identified are then examined for mathematical achievement by considering scores on the Comprehensive Test of Basic Skills (CTBS) or its equivalent. Students must score in the top 20 percent on all phases of the achievement battery. The program director then meets with students satisfying both criteria to discuss program content and student responsibilities. A student who is interested must obtain a recommendation from a mathematics teacher and submit a parent permission form. Meetings with parents are held to discuss the program and its content, materials, and logistics. This process produces a group of about one hundred fifty students out of a total middle school population of about seventeen thousand. Approximately 20 percent are in seventh grade and 80 percent in eighth grade. Typically about 90 percent of the district's middle schools are involved in the program.

Usually students are released from their mathematics classes for the honors seminar. Scheduling incompatibilities occasionally requires that students be pulled from other academic classes. The director meets with affected teachers to explain the program and to secure the release of students from class. Since the achievement levels of the students are generally high in all areas, most teachers readily agree to release the students once a week. Whether students are released from mathematics classes or from other classes, they are expected to make up all work missed during the time they attend the honors seminar. Teachers are surveyed after nine weeks to determine if their students' participation in the seminar is significantly affecting their progress in regular class.

The textbook *Unified Mathematics,* by Fehr, Fey, and Hill (Addison-Wesley), was chosen for the program because of its emphasis on the fundamental structure of mathematical systems, its integrated development of algebraic and geometric concepts, and its level of abstraction and difficulty. It also presents many topics not included in the district honors curriculum, such as mathematical groups, mathematical mappings, and the theory of binary operations and binary relations. Because *Unified Mathematics* has several volumes, it provides a wealth of material for study by students who may be involved with the program for more than one year or for a possible extension of the program into high school.

An illustration of the depth and type of approach used in the program is given by the treatment of binary operations. The students are presented with the following definitions:

1. Let # denote a way of assigning to every ordered pair of elements in a set S exactly one element of S. Then # is a binary operation, and $(S, \#)$ is an operational system.
2. An operational system $(S, \#)$ has the associative property if and only if $(a\#b)\#c = a\#(b\#c)$ for all a,b,c in the set S.
3. An operational system $(S, \#)$ has the commutative property if and only if $a\#b = b\#a$ for all a,b in the set S.
4. If an operational system $(S, \#)$ contains an element e such that $a\#e = e\#a = a$ for all a in the set S, then e is the identity element.
5. If an operational system $(S, \#)$ has an identity element e and there exist elements a,b such that $a\#b = b\#a = e$, then a and b are inverses.
6. In an operational system $(S, \#)$ if $c\#a = c\#b$ guarantees that $a = b$, then $(S, \#)$ has a left-hand cancellation law.
7. In an operational system $(S, \#)$ if $a\#c = b\#c$ guarantees that $a = b$, then $(S, \#)$ has a right-hand cancellation law.

Specific sets and operations are then selected and examined to determine which of the defined properties the resultant systems possess. For example, let S be P, the set of all points in a plane, and # be the operation Mid, which locates between each pair of points a midpoint. Students analyze the system (P, Mid) for each of the properties above. Students also create operational systems that possess any desired combination of defined properties.

Each week one or more sections of the textbook are discussed and homework assigned. Work is due the following week. No grades are issued—each assignment is rated by the instructor as outstanding, excellent, satisfactory, or poor. Since the work assigned contains many questions that require written explanations, includes many problems that can be attacked in several different ways, and encourages creative thinking, the grading is quite subjective. At the end of each nine weeks' grading period, an overall progress report is mailed directly to the parents. Contact with parents is initiated earlier if a student receives a poor rating on three assignments or if assignments are missing. If improvement does not occur, the student is dropped from the program. Students quickly determine whether they are sufficiently interested and capable. Those who are not soon find the extra work unbearable and withdraw voluntarily. The attrition rate has been about 5 percent. At the end of the school year, letters certifying the students' exceptional capabilities, their participation in the program, and their general performance level are entered into their cumulative records folder.

Students who participate in the seminar as seventh graders are eligible to continue as eighth graders. The program operates on the whole in the same way it did the first year, with students completing the first volume and continuing into the second. Occasionally, however, student ability and achievement have indicated that even more intensive mathematics is desirable. With the instructor's approval, the approval of the school, and permission from the parents, extremely able second-year students are placed in a district-wide magnet class that meets before school three days each week. Parents are responsible for transportation to and from a central location. Students placed in this class are not required to take mathematics in their schools. Students spend seventy minutes each day—twenty minutes for analysis, discussion, and questions regarding homework and fifty minutes for lecture. Students may arrange for late arrival, early dismissal, or a full schedule at their own school. Students in this class spend about four months completing volume 1 and seven of ten chapters in volume 2, about four months studying first-year algebra using the district's honors in algebra 1 curriculum, and one month studying microcomputer applications. They then enroll in honors geometry as ninth graders.

The general reaction to the program has been very positive. Student interest is high and performance is excellent. Parents have been extremely supportive. Principals and teachers have recieved the program as an enhancement to their overall offerings in mathematics. Excellent liaison has been established among individual school mathematics departments, counselors, and teachers of the gifted.

Many advocate fast-paced acceleration as the most effective vehicle for meeting the needs of students of high ability. But for gifted students acceleration is intrinsic, and only the specific format is subject to suggestion by the teacher. The approach discussed in this article provides not only for accelerated learning but also for a challenging application of very fundamental mathematics in varying contexts and for the exploration of processes of mathematical analysis and classification essential to the development of all later mathematics. As more information is gathered from experience with this group of students, refinements occur both in curriculum and in the process of identifying extremely able students. This program has provided an excellent starting point for maximizing the growth of the district's strongest mathematical talent.

22

Statistics in the Curriculum for the 1980s

Gottfried E. Noether

RECOMMENDATION 6 of NCTM's *Agenda for Action* (1980) asks for more mathematics for all students and then adds:

> To say that most students should study more mathematics is not to say that it should be the same mathematics for all. It does not mean simply keeping all students longer in the same traditional track. In fact, such a recommendation poses a tremendous challenge to curriculum developers and school districts to devise a more flexible range of options, a diversified program to meet a variety of interests, abilities, and goals. [p. 17]

No explicit reference is made to statistics as a recommended alternative, but support for statistics is strongly implied. In connection with Recommendation 2, we read:

> There should be increased emphasis on such activities as—
> - locating and processing quantitative information;
> - collecting data;
> - organizing and presenting data;
> - interpreting data;
> - drawing inferences and predicting from data. [p. 7]

That is what statistics is all about.

With this kind of encouragement from NCTM, local school districts should begin to reevaluate their mathematics offerings. There can be little doubt that statistics deserves careful consideration in any such reevaluation. The real problem is what kind of statistics.

A number of schools now offer an elective course in probability and statistics for students who have completed the traditional mathematics sequence. But more often than not this is really a course in probability *theory* with a few remarks about statistics tagged on almost as an afterthought. There is nothing wrong with a course in probability. For many high school students, probability is a much more fruitful subject than calculus. However, much useful statistics can be taught with little or no background in probability. To tie statistics to probability prevents the large majority of students from studying a subject that has ever-increasing significance for daily living.

At the school level we do not need a course in the mathematics of statistics. We need to give students hands-on experience with real data, data relating to such down-to-earth concerns of students as their favorite school subject, their favorite TV program, or their preferred color for a car. Students should be encouraged to frame their own problems, collect their own data, and try to make sense of the data they have collected. Only by being involved in the handling of data from beginning to end will students lose their mistrust of things numerical and perhaps even gain an appreciation for the statistical way of thinking. An ability to deal critically with numerical information is an increasingly important prerequisite for informed citizenship.

In recent years, I have conducted several NSF-sponsored summer workshops on statistics for high school teachers. The variety of interest groups that formed among the attending teachers is noteworthy. The approaches they suggested for teaching statistics include the following:

1. Using statistical ideas throughout the four-year mathematics curriculum, whenever suitable, to develop or illustrate mathematical concepts or simply to achieve a change of pace

2. Using statistics as a source of problems in a computer programming course

3. Using statistics as the principal component of a general mathematics course for underachievers in mathematics

4. Offering a statistics course for average and above-average students as an alternative to regular mathematics courses

Statistical Ideas That Apply Mathematical Concepts

Here is a statistical activity that fits under approaches 1 and 2 listed above and affords an exercise in coordinate plotting. For each member of last year's high school basketball team, have the students record the total number of points scored, x, and the total number of fouls committed, y, during the school basketball season. To discover a possible relationship between the number of points scored and the number of fouls committed, they plot the points in the (x, y) plane. Quite generally, players who score few points also commit few fouls and players who score many points amass more fouls. In the (x, y) plane, the points tend to scatter around a line through the origin sloping toward the upper right-hand corner. A *statistical* problem is to determine a line that is representative of the general trend. There are many different ways of finding such a line. One intuitively simple method is to rotate a ruler through the origin until an equal number of points fall above and below the ruler's edge. How can we describe this procedure in *mathematical* terms? We start with the equation $y = mx$ for a straight line through the origin. The basketball player who has scored x_i points and has committed y_i fouls furnishes the value $m_i = y_i/x_i$ (provided the point (x_i, y_i) does not coincide with the origin). Our graphical method is now seen to be equivalent to determining the median of all m_i.

Example: Table 22.1 gives the records of fifteen basketball players arranged according to the total number of points each player has scored. Figure 22.1 is a plot for the data. The two points at the origin are eliminated from further consideration. Of the remaining thirteen points, six lie above

TABLE 22.1
The Record of Fifteen Basketball Players

Player	Points (x)	Fouls (y)	$m = y/x$
A	72	28	0.39
B	62	16	0.26
C	55	17	0.31
D	50	18	0.36
E	44	21	0.48
F	31	6	0.19
G	22	7	0.32
H	20	6	0.30
I	7	3	0.43
J	6	5	0.83
K	3	0	0.00
L	1	3	3.00
M	0	1	∞
N	0	0	–
O	0	0	–

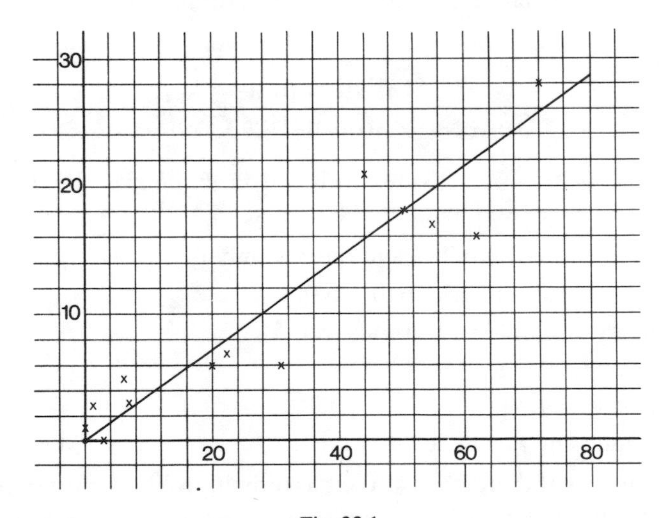

Fig. 22.1

and six lie below the line connecting the origin with the point representing player D. For player D, $m = 18/50 = 0.36$; and so we take the line $y = 0.36x$ to represent the general trend between the number of points scored and the number of fouls committed. Roughly, players commit one-third as many fouls as they score points.

Computer applications

The computer approach to this problem is relatively straightforward but requires some precautions. The computer must be instructed to disregard players for whom $x = y = 0$ and to interpret the ratio y/x as a sufficiently large positive number when $x = 0$ and $y > 0$, as for player M. The remaining programming is routine, selecting the median of all computed ratios $m = y/x$.

Fitting a general straight line $y = a + bx$ to a set of points presents a more challenging programming problem. In generalizing the method for fitting a line through the origin, we now compute slopes for all possible pairs of points (x_i, y_i) and (x_j, y_j) with $x_i < x_j$,

$$b_{ij} = \frac{y_j - y_i}{x_j - x_i},$$

and choose for the value b the median of these, b_{ij}. For the intercept, we choose the value a such that the line $y = a + bx$ has an equal number of points lying above and below the line.

Much useful material for the kind of approaches and courses listed earlier can be found in the four-volume collection *Statistics by Example* (Mosteller et al. 1973). In particular, volume 1, *Exploring Data,* is concerned with data handling—how to summarize data and how to gain insights from data through proper tabulation and graphical representation.

Statistics in a General Mathematics Course

The following sets of activities included in *Exploring Data* would be appropriate for a general mathematics course:

- Characteristics of families and their members (no. 3)
- Points and fouls in basketball (no. 4)
- Examples of graphical methods (no. 5)
- Babies and averages (no. 6)
- The cost of eating (no. 9)
- Tom Paine and Social Security (no. 13)

Readers will note that number 4, the basketball problem, proposes a solution that differs from ours. But in statistics there is seldom just one solution to a problem. This is an important lesson for persons accustomed to thinking in terms of a single correct solution to a mathematical problem.

Statistics for Average and Above-Average Students

Much of the data-handling material proposed for a statistics course for underachievers should also be included in a statistics course for average and above-average students. But students in the latter group should also be involved in *informal* discussions about drawing inferences from data. The following sections from *Exploring Data* contain useful material:

- Turning the tables (no. 10)
- Testing beer tasters (no. 11)
- Estimating the size of wildlife populations (no. 12)

Two sections from volume 2, *Weighing Chances,* are recommended:

- Random digits and some of their uses (no. 2)
- Introduction to chi-square procedures (no. 6)

Somewhat more challenging are the following sections from volume 3, *Detecting Patterns:*

- Normal probability distributions (no. 2)
- How much does a 40-pound box of bananas weigh? (no. 3)
- Grocery shopping and the central limit theorem (no. 4)
- Chi-square distributions by computer simulation (no. 5)

Helpful supplementary readings for all four types of courses are found in *Statistics: A Guide to the Unknown* (Tanur et al. 1978).

REFERENCES

Mosteller, Frederick, William H. Kruskal, Richard F. Link, Richard S. Pieters, and Gerald R. Rising, eds. *Statistics by Example.* 4 vols. Reading, Mass.: Addison-Wesley Publishing Co., 1973.

National Council of Teachers of Mathematics. *An Agenda for Action: Recommendations for School Mathematics of the 1980s.* Reston, Va.: The Council, 1980.

Tanur, Judith M., Frederick Mosteller, William H. Kruskal, Richard F. Link, Richard S. Pieters, and Gerald R. Rising, eds. *Statistics: A Guide to the Unknown.* 2d ed. San Francisco: Holden-Day, 1978.

23

New Directions for General Mathematics

Donald Chambers

A few schools may increase the time devoted to mathematics in grades K–8, but the best hope for increasing the amount of time students spend in the study of mathematics is to encourage (or require) them to enroll in mathematics courses throughout their high school program. Unfortunately, most school districts do not have a mathematics program attractive enough to entice students to enroll for additional courses on an elective basis, nor do they offer courses that would be appropriate if the requirement were increased. Before districts can require more mathematics, they must develop a curriculum that meets the needs of students who do not enroll in algebra and geometry.

The Origins of General Mathematics

High school programs typically offer four years of precalculus mathematics designed to prepare students for college. In addition they offer a collection of general mathematics, consumer mathematics, and business arithmetic courses designed for remediation in computation with whole numbers, fractions, and decimals.

School mathematics programs in the United States have always been designed with college entrance in mind. In 1745 arithmetic became an entrance requirement at Yale. Princeton and Harvard continued to teach arithmetic until 1760 and 1807, respectively, when they, too, included it among their entrance requirements. Algebra was first required for college entrance in 1820 by Harvard, and in 1865 Yale became the first college to require geometry for admission.

During the mid-nineteenth century, "grade school still concentrated on reading and arithmetic; high school, on preparation for college; and college, very frequently, on preparation for the ministry" (Rosskopf 1970, p. 6). In the twentieth century, algebra and geometry became the core of the high school mathematics program. Additional algebra, geometry, and trigonometry extended the precalculus sequence to four years, with algebra being the standard ninth-grade course.

As the population of high school students became more diverse (not all intending to enter college), the percentage of students who enrolled in algebra declined from 58 percent in 1905 to 40 percent in 1920 to 30 percent in 1934. In 1920 the Commission on the Reorganization of Secondary Education (1970) observed that the needs of "general readers," who will not continue to study mathematics or apply mathematical skills in their occupations, need to be addressed in the secondary mathematics program. In its original formulation, general mathematics was not restricted to low-achieving students in a terminal course. According to Jones and Coxford (1970, p. 51),

> The general mathematics concept was based on a reorganization of school mathematics that de-emphasized compartmentalization. In essence it was to consist of a sound, gradual development of algebra, geometry, trigonometry, and introductory statistics throughout the six years of secondary school, a development that would stress interrelationships.

In 1923 the Mathematical Association of America used the term *general courses* as a synonym for *composite, correlated,* and *unified* courses and recommended that the content of general mathematics in grades 7, 8, and 9 include arithmetic, intuitive geometry, algebra, numerical trigonometry, demonstrative geometry, history and biography, optional topics, problems, and numerical computation (National Committee on Mathematical Requirements 1970, pp. 403–10). Although several curricula that were developed during the 1960s and 1970s have adopted this unified approach (examples include the Secondary School Mathematics Curriculum Improvement Study, *Unified Modern Mathematics;* CEMREL, *Elements of Mathematics;* and the New York State *Three-Year Sequence for High School Mathematics*) and seventh- and eighth-grade programs now generally follow it to some extent, the recommendations of the MAA have not been broadly implemented in grades 9–12. "At the ninth-grade level, general mathematics was often planned and taught as a lower-level alternative to algebra rather than as the continuation of a broadly planned general mathematics program" (Jones and Coxford 1970, p. 52).

The introduction of graduation requirements in mathematics only made matters worse. Again quoting Jones and Coxford (1970)—

> A type of general mathematics not contemplated by the National Committee on Mathematical Requirements began to develop out of the increasingly common requirement that everyone take at least one year of mathematics in grades 9–12. This new general mathematics developed as the most popular alternative to algebra in the ninth grade. Down to the present day this course has been ill-defined and often poorly, or at least unwillingly, taught. [p. 53]

General Mathematics Today

In Wisconsin each school district sets its own graduation requirements. The number of districts requiring two or more years of mathematics has increased from 23 percent in 1971 to 40 percent in 1981; the number of districts requiring one year or less has decreased from 77 percent in 1971 to 56 percent in 1981.

Although more mathematics is being required, the curriculum has not been modified. NCTM's (1980, pp. 20–21) precaution that "significant mathematics courses should be available to these students in ninth and tenth grades, not just the traditional general mathematics review or prealgebra courses," has been heeded in very few instances.

Students have taken general mathematics and prealgebra courses to meet graduation requirements rather than to meet their own needs for increased mathematical competence. Large numbers of students in both the precalculus sequence and the general mathematics sequence drop out of mathematics once some specified requirement for college entrance or high school graduation has been met. Table 23.1 indicates the percent of Wisconsin students enrolled in mathematics by grade level. Precollege mathematics includes first- and second-year algebra, geometry, and other precalculus and calculus courses.

TABLE 23.1
Percent of Wisconsin Students Enrolled (or Not Enrolled) in Mathematics

Grade Level	Precollege Mathematics	Other Mathematics	No Mathematics
9	54	43	2
10	52	18	30
11	41	12	47
12	25	11	64

At the ninth-grade level, teachers (and sometimes the students themselves) determine that algebra is inappropriate for almost half the students. At the end of the ninth grade, over half the students not enrolled in algebra discontinue their formal study of mathematics. This may be because they have completed the graduation requirement, because the course was uninteresting and unsuited to their needs, or because there was no appropriate continuation course for them. Most likely it is a combination of these reasons. Although many of these students need more instruction in mathematics, most existing school programs cannot help them.

The Kenwood Conference and the New General Mathematics

In the summer of 1979 the Wisconsin Department of Public Instruction secured funding to begin developing a three-year general mathematics program for secondary schools. Sixteen Wisconsin mathematics educators met at the Kenwood Conference Center on the campus of the University of Wisconsin—Milwaukee in June 1980 to identify goals for the course and to begin identifying and developing instructional units. Zalman Usiskin of the University of Chicago met with the committee for two days and was instrumental in guiding its thinking on the applications of arithmetic.

The committee developed a list of assumptions and created a partial list of instructional objectives for the first year of the three-year course as well as some suggested activities for achieving those objectives. During the summer of 1981, a draft of objectives for the first year was completed and additional activities were written. Some, but not all, of the activities were piloted during the 1980–81 school year in general mathematics classes.

The goal of the project is not to produce student text material but to create a teacher handbook outlining the three-year program and describing each year's objectives and activities in detail. The handbook will also identify some readily available instructional resources. Although this goal will not be realized for several more years, the objectives and activities for the first year of the course are now finalized, and more attention can be focused on identifying objectives and activities for parts two and three.

The following assumptions establish the philosophy of the general mathematics program. The content objectives were developed with these assumptions in mind. However, teaching for the content objectives without continuous and conscientious attention to these assumptions is not sufficient to achieve the goals of this program.

Assumptions of the new general mathematics

1. Learning to solve problems is the principal reason for studying mathematics. The curriculum will be based on a problem-solving approach.
2. Mathematical literacy for all students is the fundamental goal of this program. The program will develop all ten basic skill areas identified by the National Council of Supervisors of Mathematics in its "Position Paper on Basic Mathematical Skills," especially the ability to solve routine problems, the ability to interpret data communicated through the mass media, and the ability to make decisions based on numerical data.
3. The National Council of Teachers of Mathematics recommends that three years of mathematics be required of all students in grades 9–12. Many Wisconsin districts are increasing their mathematics requirement. An increased requirement makes it even more important that schools provide a program of good mathematics appropriate to the needs of non-precalculus students.

4. Pencil and paper computation is not prerequisite to the mathematics that adults do, nor is pencil and paper computation more important than computation by estimation or by calculators.
5. With all its limitations, the precalculus sequence (algebra, geometry, algebra/trigonometry, analysis) is still the program which provides students the most options for further education and careers, and students with average ability and above should be encouraged to pursue the precalculus sequence.
6. The non-precalculus sequence must allow students to transfer into the precalculus sequence. It must also be possible for students in the precalculus sequence to transfer into the non-precalculus sequence.
7. Non-precalculus students need a program which provides some mathematics from each of the traditional subcategories of mathematics: arithmetic, algebra, geometry, trigonometry, statistics, probability, computers. There is no unique sequence for organizing this curriculum.
8. The curriculum should be sufficiently flexible to allow different forms of implementation in different schools.
9. Problems identified in the growing literature on mathematics anxiety must be directly confronted in this program.
10. No curriculum is teacher proof. Ultimately the success of this or any other curriculum is largely dependent on the skill of the teacher.
11. Children can and will learn mathematics if provided an appropriate program.

Objectives of the first year

> It is but fair to say that few of the specific suggestions made are in fact new, many being already somewhere actually in practice. [Commission on the Reorganization of Secondary Education 1970, p. 365]

A list of the topics or the course content identified by the Kenwood Conference fails to distinguish adequately between this concept of general mathematics and the courses currently being offered under that title. The first year devotes considerable attention to the *use* of arithmetic. Zalman Usiskin and Max Bell have cataloged the uses of numbers into six categories: counts, measures, locations in reference frameworks, scalars, codes, and nominal uses. They have also classified the uses of the operations of addition, subtraction, multiplication, division, and powering. Although virtually no attention will be given to the algorithms for performing these operations, considerable attention will be given to the uses of arithmetic and strategies for solving routine problems as identified by Usiskin and Bell.

No attempt was made to identify distinct categories and topics. Consistent with the recommendations of the National Committee on Mathematical

Requirements, topics from arithmetic, algebra, geometry, statistics, trigonometry, and computer mathematics are included. The program is organized around the following strands: uses of numbers, adjusting numbers, measurement, tables/charts/graphs, ratio/proportion/percent, similarity, formulas, probability, decision making, and problem-solving processes. These strands, together with the traditional subcategories of mathematics (such as algebra and geometry), create a matrix containing cells for which instructional activities can be identified.

The categorized list of objectives that follows identifies the content of the mathematics program for the first year. The list does not suggest the structure of the first-year course, since the topics will not be ordered consecutively for the most part. Furthermore, because of the variety of student abilities and because much of this material reviews the content commonly contained in good K–8 programs, the course will be more successful to the extent that the content can be appropriately matched with each student's background and ability.

Uses of Numbers

The student is able to—

1. distinguish between various uses of numbers (counts, measures, locations in reference frameworks, scalars, codes, nominal uses);
2. recognize the need for a variety of sets of numbers (counting numbers, negative numbers, rational numbers, even numbers, multiples or factors of a particular number);
3. order numbers (integers, rational numbers);
4. recognize many names for a given number ($16 = 2 \cdot 8 = 18 - 2 = 24 \div 1.5 = 12 + 4 = 2^4$);
5. interpret numbers used as scales (hardness, octane, etc.);
6. use numbers as locators (street addresses, latitude and longitude, number lines, Cartesian coordinates);
7. interpret formulas as another way of expressing numbers;
8. use numbers to interpret a probability statement (chance of rain, etc.).

Adjusting Numbers

The student is able to—

1. round numbers to indicated place value;
2. estimate measures;
3. estimate results of computations using rounded numbers;
4. refine estimates based on successive approximations;
5. use scales to transform numbers (one inch represents ten miles, etc.);
6. express data by graphs and other displays;
7. rewrite numbers in different forms.

Measurement

The student is able to—

1. use measuring instruments;
2. use counting techniques appropriate to measurement (count squares in an array, etc.);
3. regard measure as the repeated application of a standard unit;
4. recognize the approximate nature of measurement;
5. derive simple measurement formulas (especially area and volume);
6. apply measurement formulas;
7. use measures to the appropriate degree of accuracy;
8. interpret and use rate measures.

Tables/Charts/Graphs

The student is able to—

1. organize data into a frequency chart;
2. graph data from a frequency chart;
3. read and interpret graphs and tables;
4. use graphs to interpolate and project;
5. create and interpret coordinate graphs;
6. interpret rate as it applies to tables, charts, and graphs.

Ratio/Proportion/Percent

The student is able to—

1. express comparisons as ratios;
2. express rates as ratios;
3. interpret $\frac{b}{a} = x$ as a solution to $ax = b$;
4. find the missing term of a proportion.

Similarity

The student is able to—

1. recognize similar figures in two and three dimensions;
2. describe characteristics of similar figures;
3. determine the ratio between any two similar figures;
4. identify congruence as a special case of similarity;
5. identify the corresponding sides of similar polygons and state a proportion relating the lengths of the sides;
6. recognize maps and other scale drawings as instances of similarity;
7. use similarity to find distances that cannot be measured directly.

Formulas

The student is able to—

1. read and interpret parentheses as grouping symbols;
2. apply conventions regarding the order of operations;
3. solve formulas in which the unknown variable stands alone on one side of the equation.

Probability

The student is able to—

1. express experimental probabilities using the ratio of successes to total trials;
2. list the possible outcomes given a description of an experiment;
3. estimate the likelihood of an event given a description of a simple experiment;
4. estimate the probability of an event based on the possible outcomes and their stated probabilities or data that enable those probabilities to be calculated.

Decision Making

The student is able to—

1. make decisions based on statistical data or probabilities;
2. locate and run preexisting computer programs;
3. participate in simulations that emphasize decision making;
4. interpret and use data from computer output in making decisions or solving problems;
5. recognize common valid and invalid inference patterns.

Problem-solving Processes

The student is able to—

1. organize data into pictures, tables, charts, or graphs;
2. identify problem constraints;
3. identify a related problem or a simpler problem;
4. use standard problem-solving strategies (working backward, patterns, guess and test, etc.);
5. check a solution against problem constraints;
6. find alternative solutions or solution patterns;
7. create a situation having a particular mathematical formulation;
8. use language compatible with a particular computer system in order to perform a specific task;
9. type and run a preexisting computer program;
10. modify a computer program to produce different output.

The structure of the first year

General mathematics courses almost always begin with an extensive review of whole-number arithmetic. This review, along with a review of fractions and decimals, frequently expands to compose the entire course.

The Kenwood program begins with an orientation to four-function calculators and immediately exposes students to the breadth of topics in the course through problem-solving activities from each of the areas listed previously. This unit of work, which ordinarily does not exceed two weeks, exposes the students to problem solving, the uses of arithmetic, and the use of the calculator as a problem-solving tool as well as the scope of the first-year course.

The arrangement of topics beyond this first unit is very flexible. The topics are not intended to serve as units of instruction and are not to be treated in isolation from each other. The stressing of interrelationships identified as a goal in the 1920 report of the Commission on the Reorganization of Secondary Education (1970) and the unified approach recommended in the 1923 MAA report, *The Reorganization of Mathematics in Secondary Education* (National Committee on Mathematical Requirements 1970) are fundamental to the structure of each year of this three-year course. The activities reported in a later section of this article illustrate this interrelationship of topics.

In order to give the appropriate emphasis to problem solving, problem-solving techniques are taught, but within the context of the strands. In addition, each class period is to be organized around some problem-solving activity consistent with the first recommendation of *An Agenda for Action.*

The second and third years

The same strands identified for the first year serve as organizing elements for the second and third years. As with any curriculum based on a spiral approach, both the second and third years review material developed during the previous year. Furthermore, it is no less true in the second and third parts than in the first that students enter with diverse backgrounds and abilities. Many students will not complete all the objectives of the first year, and the second year's course must be adjusted to provide instruction in those first-year objectives that some students missed or did not sufficiently master during the first year.

Applications of arithmetic will continue to receive emphasis. In addition, applications of algebra and geometry will be developed. The following objectives from the similarity strand are offered as examples of second- and third-year objectives.

Similarity

The student is able to—

1. draw a polygon given sufficient information regarding sizes of angles and lengths of sides;
2. estimate the lengths of one side of two similar polygons using the figure and the proportion;
3. draw similar polygons given the relationship between the sides;
4. enlarge or reduce the size of a polygon given the ratio between the two;
5. identify and correctly name similar polygons from various drawings;
6. make drawings illustrating an applied problem;
7. recognize the constant ratios between corresponding sides of similar right triangles;
8. correctly identify the ratios associated with sine, cosine, and tangent;
9. read trigonometric tables;
10. use trigonometric tables and calculators to solve simple trigonometry equations;
11. solve simple problems using applications of right-triangle trigonometry.

The first year includes very modest objectives for ratio/proportion/percent. This is not because these topics are unimportant; they are among the most important. But since they are among the most difficult in the program being proposed, they will be developed more thoroughly in the second and third years of the program.

Sample activities

The following activities from the first year's course illustrate the interrelationship of strands. The strand *scales* is related to *uses of numbers, adjusting numbers, measurement, ratio/proportion/percent*, *formulas,* and *tables/charts/graphs.*

Activity 1: Find ten different uses of numbers in the local newspaper. Discuss which are scales. Find five new uses of numbers that are scales in the newspaper. See if the scales can be categorized by type. (Uses of Numbers)

Activity 2: Without being restricted to the newspaper, find other examples of scales; for example, attitude scales, meat (rare to well done), Richter scale, insulation value, energy efficiency of appliances. (Uses of Numbers)

Activity 3: Discuss the "top 40" as a scale and changes in the order of particular tunes from week to week. (Adjusting Numbers)

Activity 4: Rank occupations according to salary. (Adjusting Numbers)

Activity 5: Transform values from one scale to another; for example, temperatures from Celsius to Fahrenheit. (Adjusting Numbers)

Activity 6: Use the resequencing function on a computer to relabel the steps in a program. (Adjusting Numbers)

Activity 7: Discuss the use of rulers and other measuring instruments as scales. (Measurement)

Activity 8: Use scales in map reading (find cities, use mileage numbers, use legend to compute distance, use rulers to measure scale and distance, use mileage charts, use a time scale for distance). (Measurement)

Activity 9: Identify ratios that are used as scales (mph, rpm). (Ratio/Proportion/Percent)

Activity 10: On a map use map distance and indicated mileage to determine the map scale. (Ratio/Proportion/Percent)

Activity 11: Use conversion formulas as changes of scale. (Formulas)

Activity 12: Draw a graph that expresses a scale (temperature to calories). Note the role of slope in drawing comparisons. (Tables/Charts/Graphs)

Activity 13: Observe the effects of choices of scale on the appearance of graphs. Determine the most appropriate choice. (Tables/Charts/Graphs)

Alternatives

Most secondary teaching of mathematics is based on the textbook. The changes recommended here have limited chance of widespread implementation until commercial textbooks consistent with the philosophy and content of the Kenwood Conference program become available. Meanwhile, it would be a step in the right direction if teachers would look at their current algebra and geometry texts, select the most relevant portions of them, and teach them from a problem-solving point of view. A very respectable and valuable one-year course could result. Students could also profit from a one-semester course in intuitive statistics and a one-semester computer course.

The traditional role of mathematics as a filter must be abandoned. We cannot sort students according to whether they can or cannot succeed in ninth-grade algebra and interpret that to mean that they can or cannot succeed in mathematics. As Paul Trafton (1980) wrote:

> The times in which we live require that all students master as much mathematics as possible. We cannot afford to have large numbers of people who have only a severely limited proficiency in mathematics. While low achievers and programs for low achievers pose difficult problems, we must apply what we know about effective instruction and sound learning. Only to claim that we did what we could while students who can learn fail to learn is not acceptable. [p. 21]

REFERENCES

Commission on the Reorganization of Secondary Education. *The Problem of Mathematics in Secondary Education,* 1920. Reprinted in *Readings in the History of Mathematics Education,* edited by James K. Bidwell and Robert G. Clason. Washington, D.C.: National Council of Teachers of Mathematics, 1970.

Jones, Phillip S., and Arthur F. Coxford, Jr. "Mathematics in the Evolving Schools." In *A History of Mathematics Education in the United States and Canada,* Thirty-second Yearbook of the National Council of Teachers of Mathematics, edited by Phillip S. Jones and Arthur F. Coxford, Jr. Washington, D.C.: The Council, 1970.

National Committee on Mathematical Requirements. *The Reorganization of Mathematics in Secondary Education,* 1923. Reprinted in *Readings in the History of Mathematics Education,* edited by James K. Bidwell and Robert G. Clason. Washington, D.C.: National Council of Teachers of Mathematics, 1970.

National Council of Teachers of Mathematics. *An Agenda for Action: Recommendations for School Mathematics of the 1980s.* Reston, Va.: The Council, 1980.

Rosskopf, Myron F. "Mathematics Education: Historical Perspectives." In *The Teaching of Secondary School Mathematics,* Thirty-third Yearbook of the National Council of Teachers of Mathematics, edited by Myron F. Rosskopf. Washington, D.C.: The Council, 1970.

Trafton, Paul R. "Assessing the Mathematics Curriculum Today." In *Selected Issues in Mathematics Education,* edited by Mary Montgomery Lindquist. Chicago: National Society for the Study of Education; Reston, Va.: National Council of Teachers of Mathematics, 1980.

Recommendation 7

MATHEMATICS TEACHERS MUST DEMAND OF THEMSELVES AND THEIR COLLEAGUES A HIGH LEVEL OF PROFESSIONALISM

24

Problem-solving Activities for Prospective Elementary School Teachers

Robert G. Marcucci

THE recommendations for school mathematics proposed by the National Council of Teachers of Mathematics (1980*a*) stress problem solving as the focus of mathematics instruction. One significant implication of these recommendations concerns the training of prospective mathematics teachers in our colleges and universities. It seems clear that if the goals of the NCTM are to be achieved, our present and future mathematics teachers will need to broaden and develop their problem-solving skills. In addition, they will need to become effective teachers of problem solving. This article describes several activities that have been used successfully in mathematics methods and content courses at San Francisco State University to strengthen the problem-solving skills and instructional techniques of prospective elementary school teachers.

Activity 1: Collecting Problems

Perhaps one of the most overlooked components of problem solving in elementary school mathematics is the problems themselves. Although the usual verbal problems of the type found in textbooks provide an opportunity to apply previously learned concepts and skills, they should be supplemented with more challenging nonroutine problems and real-world applications. This gives children the opportunity to learn and practice a variety of problem-solving strategies. Furthermore, interesting problems usually motivate children (and their teachers) to become active problem solvers.

Because a collection of good problems can enhance problem-solving instruction, our prospective elementary school teachers are asked to assemble a "problem book" during the semester. The book is composed of three sections: problems of the type found in textbooks, media and theme problems, and nonroutine problems, puzzles, and strategy games.

Textbook problems

These are written by our students in the manner of those found in textbooks. They call for the application of number and of geometric and measurement concepts and skills. The problems must be lively and of varying difficulty (such as one-step or multistep problems), describe a setting familiar to children, and be appropriate for specific grade levels (K–3, 4–6, for instance).

Media and theme problems

Articles from newspapers and magazines are an excellent source for problems, as are interesting themes or activities (such as planning a picnic). Each article and theme must be accompanied by several problems and the type of information that will be needed to solve each problem. Many of these problems encourage the use of a calculator. One of our students suggested the following problems based on the activity of planning a picnic and the information that would be needed to solve them.

1. How much should we spend on food? (number of people; price of specific foods such as sandwich makings, chips, drinks, etc.; amount each person is expected to eat)
2. How many cars will we need to get to the picnic? (number of people invited; number of people that each car will hold)
3. How much will we spend on gasoline to get to the picnic? (distance to the picnic, miles each car gets per gallon, the price of gasoline)

Nonroutine problems, puzzles, and strategy games

Finding problems for this section usually requires our students to do some research in the curriculum library or resource center. The problems and puzzles must be accompanied by detailed solutions, and the directions for playing strategy games must be clearly stated. Some games are played during class in order to discover a winning strategy.

Toward the end of the semester students submit their favorite problem from each section to the instructor. These problems are collected in a "super problem book," and each student receives a copy on the final day of class. (A sample collection of such problems, unclassified by type or level, can be found at the end of this article.)

Activity 2: Strategies

Many prospective elementary school teachers are unaware of the most basic problem-solving strategies. Over the course of the semester we devote approximately three weeks to teaching general problem-solving strategies, emphasizing the ways they can be used to attack and solve problems. Some of the strategies are drawing a diagram, simplifying the problem by using smaller numbers, guess and test, making an organized list or table, acting out the problem, working backward, using a calculator, using a formula (for example, $A = lw$ for the area of a rectangle), and using an equation. A detailed description of these and other problem-solving strategies can be found in Polya (1957), Krulik and Rudnik (1980), and the NCTM 1980 Yearbook, *Problem Solving in School Mathematics* (1980*b*).

To develop familiarity with these strategies, our students solve a variety of assigned problems in as many ways as they can, clearly indicating which strategies were employed. As an illustration, consider one student's solutions of the following problem.

Problem: Joe's mom asked him to cut a 24-inch board into two pieces. He cut one piece 6 inches longer than the other piece. How long is each piece?

Solutions

- Drawing a diagram; using an equation:

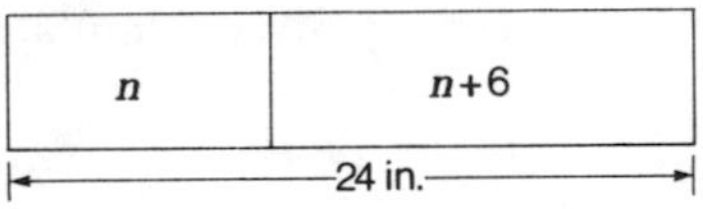

n	$n+6$

←——24 in.——→

shorter piece + longer piece = 24 inches

$$n + (n + 6) = 24$$
$$2n + 6 = 24$$
$$2n = 18$$
$$n = 9 \text{ inches (shorter piece)}$$
$$n + 6 = 15 \text{ inches (longer piece)}$$

- Guess-and-test; making an organized list:

Guess		Test
Shorter piece	Longer piece	Total
6 inches	12 inches	18 inches
10	16	26
8	14	22
9	15	24 (solution)

- Acting out the problem: Strips of paper 24 inches long are marked at 1-inch intervals. A strip is cut, and the lengths of the two pieces compared. If one piece is not 6 inches longer than the other, then another strip is cut and the lengths of the two pieces compared. This continues until a solution is obtained.

- Working backward; drawing a diagram: One piece is 6 inches longer than the other, so cutting 6 inches off the longer piece makes both pieces the same length (see diagram below). The two equal pieces have a total length of 24 − 6 = 18 inches; so each of these pieces must be 9 inches long. It follows that one piece is 9 inches and the other piece is 9 + 6 = 15 inches.

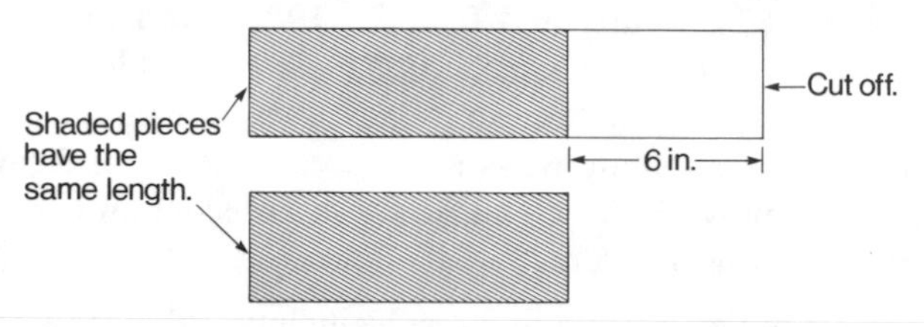

Activity 3: Modifying Word Problems

There are several ways that verbal textbook problems can be modified to teach problem solving effectively. A few simple techniques are especially helpful in the classroom because they often stimulate children's interest in solving problems.

Four modifications of a textbook type of word problem are illustrated below. In addition, several classroom questions are suggested for each modification. The original problem, its modifications, and some suggested questions were proposed by one of our students.

Problem: On her tenth birthday Anne's grandparents gave her $20. If Anne added $1 from her allowance to this money each week, how much money would she have on her eleventh birthday? (Assume 52 weeks in one year.)

- A problem without numbers: On her tenth birthday Anne's grandparents gave her some money. If Anne added some of her allowance to this money each week, how much would she have on her eleventh birthday? What is an allowance? What would we do to solve this problem? What information do we need?
- A problem without a question: On her tenth birthday Anne's grandparents gave her $20. Anne added $1 from her allowance to this money every week. What questions could we answer using this information?
- A calculator problem: On her tenth birthday Anne's grandparents gave her $20. If Anne added $1.75 from her allowance to this money each week, how much money would she have on her eighteenth birthday? How much money would Anne add to the $20 in one year? Let's

use a calculator to find out! How many years does Anne add money to the $20?

- A related problem: On her tenth birthday Anne's grandparents gave her $20. If Anne added $1 from her allowance to this money each week, how long would it take her to save enough money to buy a $200 stereo system? What do we want to find in this problem? Do we know a problem that is similar to this one? How did we solve the other problem? Can we solve this problem the same way?

Summary

The problem-solving activities described in this article can be easily incorporated into most mathematics methods and content courses for prospective elementary school teachers in our colleges and universities. These activities are designed to improve problem-solving skills and to suggest effective classroom instruction. In particular, they can be used to generate a wealth of interesting and challenging problems, teach problem-solving strategies and their applications, and motivate children to solve problems. An overwhelming majority of our students at San Francisco State University have responded positively and enthusiastically to these activities.

REFERENCES

Krulik, Stephen, and Jesse A. Rudnick. *Problem Solving: A Handbook for Teachers.* Boston: Allyn & Bacon, 1980.

National Council of Teachers of Mathematics. *An Agenda for Action: Recommendations for School Mathematics of the 1980s.* Reston, Va.: The Council, 1980*a*.

———. *Problem Solving in School Mathematics.* 1980 Yearbook. Edited by Stephen Krulik. Reston, Va.: The Council, 1980*b*.

Polya, George. *How to Solve It.* 2d ed. Princeton, N.J.: Princeton University Press, 1957.

Problem Book

Here is a sample of problems written or collected by our student teachers. The problems have not been classified by type or grade level.

- If we give each boy at a picnic three apples from the basket, one of the boys would get only two apples. But if we give each boy two apples, eight apples would remain in the basket. How many apples are there?
- You are the organizer of a tennis tournament. Twelve people will play in the tournament. Each person must play one match with each other person. How many matches must you schedule?

- How many triangles can you count in this cat?

- There are 23 children in our class, 27 children in the fourth grade, and 21 children in the fifth grade. If our class and the fourth grade are in the library at the same time, how many children will be in the library?

- A substitute teacher walks into her classroom and looks at her seating chart. She sees the following:

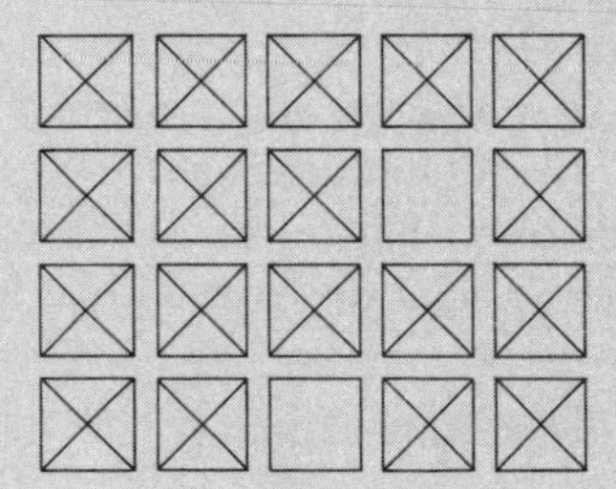

She decides the 's mean the desk is occupied and the blanks ( ) mean there is no one sitting there. Since the class is going on a field trip (and of course no one will be absent on that day!), the teacher wants each student to have a partner. How many pairs will there be?

- The elephants at the circus eat a total of 130 pounds of food every day. There are 5 elephants. How many pounds of food does each elephant eat?

- John came up to bat 6 times in the baseball game. He made 1 double, 2 singles, 2 triples, and a home run. How many bases did he touch by the end of the game?

- The Olive family went to the movies. The mother and father's tickets cost \$7.50. Their three children's tickets cost \$1.50 each. Mr. Olive gave the cashier \$20.00. How much change did he get back?

- Julie has a weekly allowance of \$3.75. She is saving to purchase a bike. Her parents have agreed to pay her \$2.25 an hour for doing odd jobs around the house. At the end of 4 weeks Julie has earned \$87.00 altogether. She worked an equal number of hours each week. How many hours did she work each week?

- Joe was going to shop and prepare dinner for his mother. He planned the menu by looking at the ads in the newspaper. He bought three lamb chops at 97¢ apiece. Corn was on sale at four for $1.00. He bought two ears. He bought one pound of butter for $2.00. Joe also purchased salad makings at these prices:

  Iceberg lettuce—1 head, 59¢
  Romaine lettuce—1 head, 69¢
  Radishes—2 bunches, 29¢
  Scallions—3 bunches, 89¢
  Tomatoes—6 for $1.00

  Joe's mother gave him $10.00. How much did the groceries cost altogether? How much change did he have left?

Number Search

- Harry the horse wants 12 apples for his lunch, but before he can have them, he must find the number **12** twenty-one times. Can you help Harry earn his apples? Solve the puzzle below. To find the number 12, add two or more numbers. You may add across, up and down, or diagonally. One is done for you.

1	6	3	7	4	3	5
4	8	9	8	6	4	4
4	3	7	3	3	3	9
3	5	9	1	4	8	1
6	6	8	2	6	8	3
8	5	2	5	1	7	5

Solution

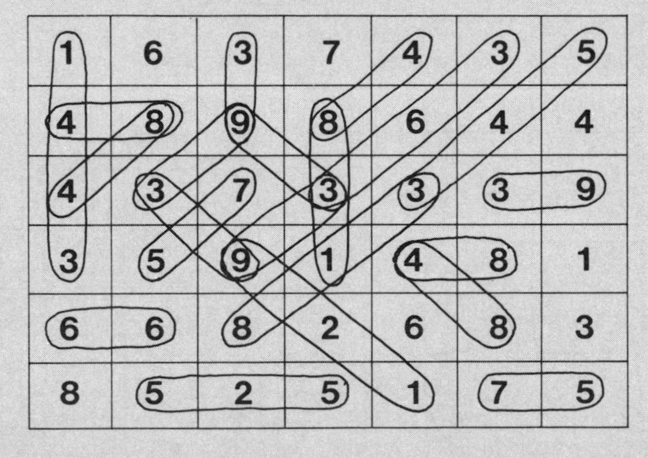

Fraction Trek

- *You will need:*
  Two markers
  A die

Two can play.

To get ready:
Cut out scorecards. Each player needs one scorecard.

How to play:
Players put their markers on START.
Players take turns. Each time it is your turn, do this:
Roll the die. Move your marker the number of spaces shown on it.
Move in any direction and on any path.
Do the problem in the space you land on. Make sure your answer is in the simplest form. Shade in that fraction on your scorecard.
The first player to complete a scorecard is the winner.

Note: Skills needed: Adding and subtracting like fractions less than or equal to 1; renaming fractions in simplest form.

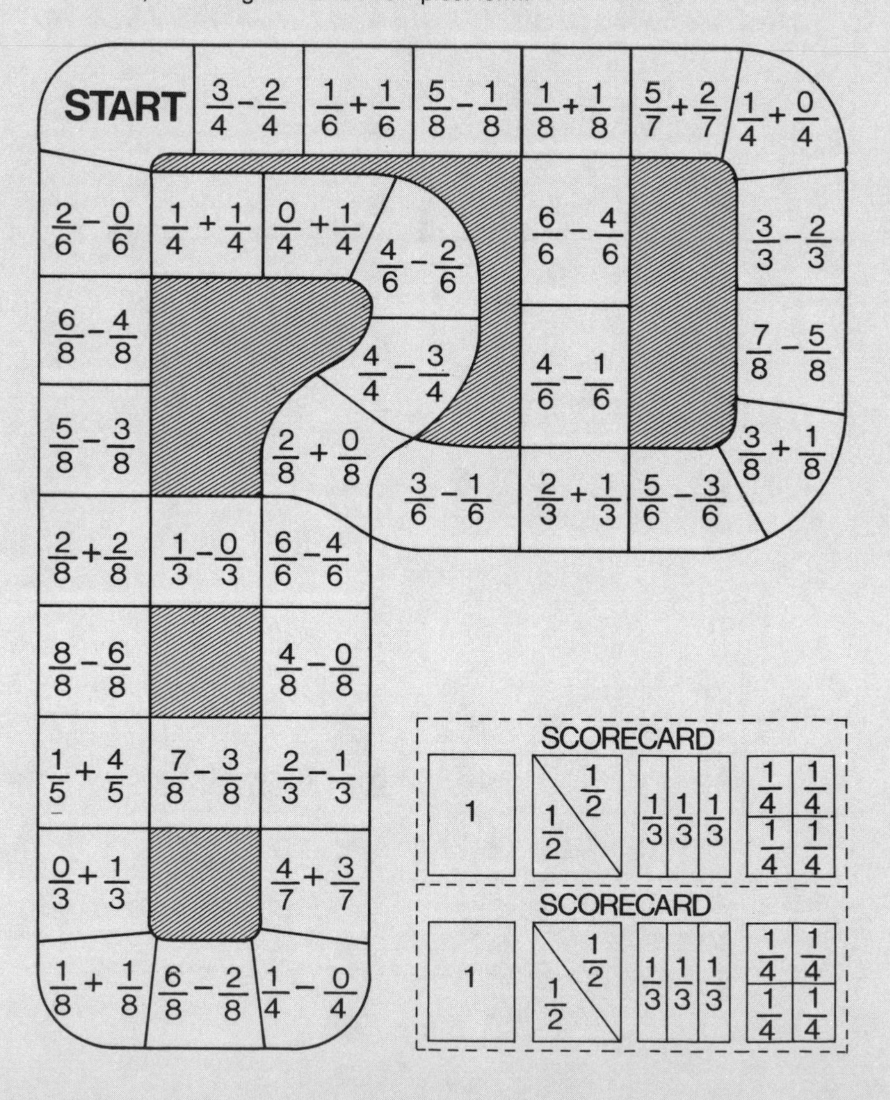

25

A Problem-solving Proposal for the Preservice Elementary School Teacher

Lowell Leake

THE FOCUS on problem solving and the inclusion of calculators and computers in the mathematics curriculum of the 1980s will certainly require that arithmetic teachers be trained in the techniques and methods of teaching problem solving and in the use of calculators and computers. One good way to provide this training for future elementary school teachers is in the mathematics content course they take as undergraduates. This course is usually called something like "mathematics for elementary school teachers" and typically lasts for one academic year. Historically, the course is one of the important results of the work of the Committee on the Undergraduate Program in Mathematics (CUPM 1961) of the Mathematical Association of America, whose efforts took place largely in the late 1950s and early 1960s. Generally, the content course is followed by a much shorter experience in the methods of teaching elementary school mathematics.

This article describes some successful experiments used by the author to incorporate the recommendations of the National Council of Teachers of Mathematics (1980) and the National Council of Supervisors of Mathematics (1977) into the philosophy of the content course. If it is desired that elementary school teachers teach problem solving, use calculators intelligently in their classrooms, and, generally, teach mathematics in accordance with the NCTM recommendations, they must be taught mathematics by college instructors who illustrate, *by example,* the goals of the 1980s. If the future teachers are themselves taught this way, they are much more likely to be successful teachers of problem solving in their own classrooms. The ideas presented here can easily be extended to in-service training for teachers already in the classrooms and they can be adapted for inclusion in the methods course, Here are some examples of how problem solving can become the focus of the content course in the preservice training of the elementary school teacher.

First Example

Perhaps the best place to begin is with a practical application of elementary mathematics that any student will recognize as important in business or home affairs. For example, present the following problem to the class. Remember, everyone has a calculator and has had some instruction and practice in using it.

> **The L-shaped (or H-shaped, etc.) hallway of the floor where you are taking this class is to be carpeted. How much carpeting is needed and how much toe strip is needed to fasten the carpet along the edges of the walls? How much will it cost?**

Suggest that the class form teams of three or four members to solve the problem and ask each team to discuss among themselves the elements of the problem. Is there enough information? Are there unstated assumptions? Can they use calculators? Is this a problem that anyone might meet at home or on the job?

Then, without telling them how to do it, ask the teams to solve the problem, using any available resources, such as the textbook. Be sure that appropriate measuring devices are available. Chances are that one class period may be inadequate; therefore, allow enough time or ask them to finish on their own as an assignment.

Here are some considerations that should come up in a very natural way among the students during the planning and action stages of solving this problem.

1. Consider the measurement process—units, possible error, precision.
2. Discuss area versus perimeter.
3. Apply area formulas to regions that are combinations of rectangles, such as L-shapes or H-shapes.
4. Consider perimeter formulas. What about doorways—do they get toe strip? What do they get?
5. How about missing information? Since price is not given, can a graph or chart be used to show the total cost of carpet based on the relationship of one price per square yard to different areas? Can a chart be constructed to show total cost of carpet based on a wide variety of prices per square yard and total areas?
6. Is labor included in prices? Is padding? What about waste and the fact that carpet comes in certain standard widths?
7. Can the budget afford the cost? What budget?
8. How can you estimate an answer? Is the solution reasonable?

This approach with this problem does several things. First, it involves the students, actively and by example, in a genuine problem. By working through the problem, helping each other, asking the instructor for help or suggestions when stumped, and solving other problems, they will know what Polya (1957) meant when he wrote about the steps in solving a problem. Polya stated that one must first understand the problem, then develop a plan, carry out the plan, and finally look back. Generally, memorizing Polya's list of steps for solving a problem (or anyone else's list) is useless. If, however, one solves several problems and thinks back about the process while reading Polya's list of steps, the real meaning of the steps can hardly be avoided. The only way to become a good problem solver is to solve problems—no one can become a good swimmer without going swimming.

Second, the carpet problem presents a situation that shows the interrelationships of many parts of mathematics—in this instance, arithmetic, geometry, measurement, graphing, and the organization of data. The future teacher will realize that these topics are not separate units or chapters.

Third, the approach involves the future teacher, by example, in an activity or laboratory setting. There is little lecturing or textbook reading.

Fourth, the problem is obviously a practical application that might be used by anyone in a home situation or by someone in the carpet installation business.

Fifth, the problem immerses the future teacher in at least eight of the ten basic skills recommended by the National Council of Supervisors of Mathematics (1977). Only probability and computer literacy seem to be missing, and calculation is clearly only one part of a much larger situation and one that the calculator can easily handle.

Variations and extensions

It has been assumed that the hallway is L-shaped or H-shaped—that is, two or more rectangles stuck together in some way. Clearly the problem can be simplified or complicated by changing the shape of the region to be carpeted. A triangular shape, for example, would show the need for an area formula for triangles. After the students have solved the L-shaped hall problem in an activity setting, they could be presented with some hypothetical problems in the classroom where the area to be covered was a triangle, an irregular polygon, or even a circle. This would be a natural way to lead into a more traditional study of the areas of plane figures, and few students will think, "What good is this?"

Tiling the floor may be preferable. Tiles usually come in one-foot squares and forty-five to the box. This adds another factor to the problem: How many boxes should be ordered? Can a fraction of a box be ordered?

Another idea would be to paint the classroom or wallpaper a room in a home. How many square feet of wall does one gallon of paint cover or one roll of wallpaper?

Another suggestion is to ask the teams to prepare bids for a contract to carpet or tile the floor or paint the walls of a room. Metric units can be used; in England, for example, some carpet comes in squares that are 50 cm by 50 cm.

Second Example

The idea for this example came from two articles in the *Mathematics Teacher* (Lappan and Winter 1980; Simon and Holmes 1969). This was done deliberately to demonstrate that if the college instructor lacks specific ideas for this kind of problem-solving approach, a little reading in the journals or other literature can fill the gaps. Here's the problem:

> **A basketball player who shoots 70 percent at the free-throw line is faced with the "one and one" situation. This means that if the player makes the first shot, another shot is awarded. Therefore, it is possible to make 0, 1, or 2 points on the play by missing the first shot (0 points), making the first and missing the second (1 point), or making both (2 points). Which is most likely to happen? Over the long run, what percentage of the time will the player make 0, 1, or 2 points?**

This problem is quite different from the previous one. For most future teachers it does not appear to have any practical application whatsoever, and it is certainly not the kind of problem one faces in everyday life. Furthermore, it involves much more sophisticated mathematics and reasoning than the carpet problem. Finally, it cannot be assumed that the students have already had some earlier experience in studying the necessary mathematics. Nevertheless, it is a good vehicle for teaching problem solving and illustrating activity learning. Furthermore, it illustrates beautifully what research (Piaget and Inhelder 1975) says is the appropriate approach for teaching probability to students who may not yet have reached the formal operational level in their cognitive development. A word of caution: The problem posed is certainly not a starting point for the elementary school teacher to study probability in mathematics, and it is assumed that some groundwork has been laid (by solving simpler problems, of course).

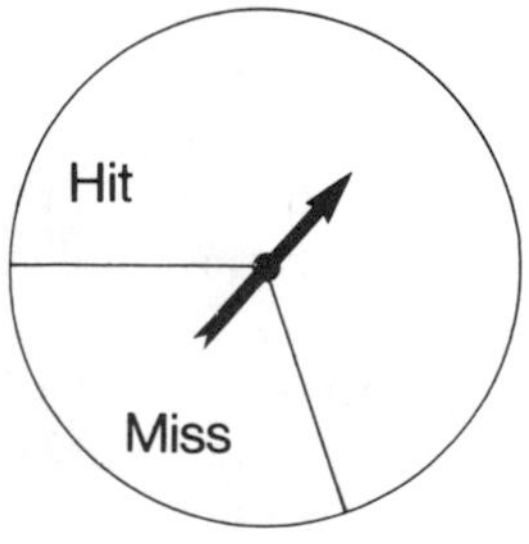

Briefly, the solution proposed by Lappan and Winter is based on a Monte Carlo approach, not on formal probability through permutations and combinations. They suggest using a spinner, such as the one shown, divided into two regions where 70 percent of the circular region represents making the shot and 30 percent represents missing the shot. Notice that this introduces simulation and

Hit

Miss

modeling in a natural way and will lead quickly to the use of computers. A trial consists of spinning the spinner once and then a second time if the first spin represents making the shot. Many trials are conducted and records kept of how many of the trials represent 0, 1, or 2 points. For example, table 25.1 shows the possible results of 200 trials. Finally, the ratios of the results to the number of trials are computed and converted to percents.

TABLE 25.1
Possible Results of 200 Spinner Trials

Points	Number of Times	Percent
0	110	55
1	36	18
2	54	27

Once again, the class can be divided into teams of three or four each. Remember, it is assumed that they already have some experience with simpler problems, such as estimating the probability of getting three heads when five coins are tossed using cards, dice, and even random digits for the simulation.

The first step is to make sure that they understand the problem. Then allow them the opportunity to develop a plan. Certainly, during this stage considerable discussion should occur, and each student should be encouraged to estimate the answer. Teams that have plans should go ahead and carry them out. The spinner idea may not even come up unless they have used spinners before or the teacher has suggested them. Blank spinners should be provided if this technique is used.

Consider what this probability problem can do for the future teachers. First, it shows emphatically that more than one method can be used to find a correct answer. Second, the problem develops a strong sense of the concept of probability from a concrete situation. Third, it points out the use of tables to record data—data from which computations are made using a calculator and from which the relationships between decimals and percents are computed. Fourth, the links between statistics and probability are shown. Fifth, in the case of the spinner, students will have to divide the circular region into the appropriate regions, probably using protractors. This will involve them in degree measurement and percents, and it will again show the need for geometric interpretation of a mathematical situation. Sixth, the approach demonstrates the kind of methodology that the future teachers may eventually use in their own classrooms to teach their students simple probability concepts. Seventh, this kind of problem illustrates beautifully the meaning of simulation and mathematical modeling, and it provides an ideal entry for the use and need of computers to produce large numbers of trials. It would be difficult to think of a better way to lead into the subject of computers. If this opportunity is pursued, the future teachers will see that the problem that appeared to have no practical application illustrates a method that is of vast importance to applied mathematics and that this idea can be passed on to elementary school children in the upper grades. How many opportunities are there to show genuine applications of mathematics in the content course for elementary school teachers?

A word of caution is in order here. The estimation of the answer in this kind of problem may be an out-and-out guess and could be quite different from an estimation for the carpet problem, where one might pace out the dimensions. Probability problems are notorious for having nonintuitive results; the well-known birthday problem is an excellent example (Kemeny, Snell, and Thompson 1957). It is probably quite healthy for students to see in practice the difference between a guess, even an intelligent one, and an informed estimate.

Variations and extensions

Variations on this problem would include the consideration of players whose free-throw shooting percentages are, for example, 60 percent or 80 percent. Simon and Holmes (1969) use this paraphrased version of a problem concerning basketball:

> **A 60 percent free-throw shooter decided to change her style of shooting. After practicing the new style for some time, she decides to test it out. She shoots ten free throws and makes nine of them. Is this sufficient evidence to claim that she is now a better free-throw shooter than before?**

This is a much more sophisticated question, since it introduces the ideas of hypothesis testing and level of significance. However, these are very important ideas in the application of mathematics and deserve consideration. A trial in this problem would be to shoot ten shots (using spinners or random digits or computers) and see if nine or ten shots are made. The null hypothesis is that the player still shoots 60 percent. Therefore, the spinner should be divided on a 60-40 basis. After a large number of trials it may become clear that someone who shoots 60 percent will occasionally make nine out of ten or even ten out of ten. The question is, how rare are these occasions? If they are not very rare, say more than 5 percent of the time, then the evidence is not sufficient for claiming an improved shooting percentage—chance might explain the performance. This version of the problem leads easily into the need and use of statistical methods—for instance, in medical research. How does one decide whether or not a new vaccine works, for example?

Third Example

Back to Basics Can Drive You Buggy

Another recent article in *Mathematics Teaching* (Barclay 1980) uses an old idea in a more elaborate and effective way. Barclay described what happened in a middle school classroom, but the idea can be used to great advantage in a college course in mathematics for elementary school teachers. The preservice teachers are presented with several examples of children's incorrect algorithms and are asked to diagnose the students' errors. For each incorrect algorithm there should be two or three examples. Proof of the correct diagnosis is provided by doing some additional problems exhibiting the same error. For example, what is wrong with the following addition problems?

$$\begin{array}{r} 28 \\ +\ 42 \\ \hline 610 \end{array} \qquad \begin{array}{r} 954 \\ +\ 581 \\ \hline 14135 \end{array}$$

What would you get if you did these the same way?
a) 278 + 395 *b)* 234 + 125
Why does the incorrect algorithm work on *b*? And so on.

First stage

Break the class into teams of three or four again. Present each team with a ditto sheet of several examples of incorrect algorithms. It is probably best to stick to addition, subtraction, and multiplication at first. Ask them to work together within their teams to diagnose the errors. Let them organize themselves any way they want. It might be a good idea to let the teams compete with each other on a timed basis.

Second stage

Have each team member invent her or his own incorrect algorithm, with examples. In turn, each individual challenges the teammates to diagnose the incorrect algorithm. When each team finishes, have them select their "best" incorrect algorithm to present to the other teams for their diagnosis.

What are the bonuses of this problem situation? First, the diagnosis of incorrect algorithms is an important skill for any teacher; it is not exclusively a topic for a methods course, since genuine mathematical reasoning is involved and diagnosing incorrect algorithms is not a trivial matter. Second, it is almost inevitable that diagnosing incorrect algorithms will lead to better understanding of correct algorithms and, it is hoped, that some curiosity about the nature of algorithms will emerge. If these future teachers now return to a classroom discussion of algorithms, there will be a much better chance that they will appreciate an analysis of them.

Variations and extensions

This idea presents a golden opportunity for the future teacher if it is extended to calculator problems. For example, ask them what student A is doing wrong with the calculator if he or she gets the answer −0.935 to three places for the problem

$$\frac{7.2 + 13.8}{9.7 - 3.1}$$

(correct answer to three places is 3.182). Or, why does calculator X get 16 for $3 + 5 \times 2$ and calculator Y get 13? These questions immediately force students to think about what their calculators do. Do they use arithmetic or algebraic logic? It also makes them think of the implied order of operations for the symbols used and the various techniques for using calculators; in short, a different class of algorithms is introduced. The use of memories, the necessity for writing down intermediate results, and overflow are examples of related problems that can be introduced and pursued. What about Reverse Polish Notation (RPN)? Many of the problem-solving techniques and the "ten basic skills" are involved in figuring out how to use a calculator correctly—for example, estimation, reasonableness of results, choice of operation, and so on.

The college instructor must realize that the particular problems discussed above are not the key idea. What matters is that the future teachers will be immersed in a problem-solving situation carefully selected by the instructor and that when they become teachers, there will be a much better chance that they will at least attempt to teach their students how to solve problems. These teachers will know what problem solving means, and they will have experienced problem solving themselves. They will be much better equipped to teach the mathematics of the 1980s than teachers who experience a more traditional approach in the content course.

How to Get Started

For a college instructor who wants to try this approach in the content course, a few suggestions will be given. First, do some reading about problem solving. The 1980 Yearbook of the NCTM is an excellent place to start; it is filled with good ideas that can be used in the college classroom (NCTM 1980). It might even be a good idea to have the future teachers in the content course buy the 1980 Yearbook as a supplementary text. Another excellent choice for a supplementary test would be *Problem Solving—a Basic Mathematics Goal* (two paperback booklets), edited by Gibney, Pikaart, and Suydam (1980). A third possibility is *Problem Solving: A Handbook for Teachers,* by Krulik and Rudnick (1981). It is probably best to begin problem solving early in the year, perhaps even on the first day of class. Let the future teachers know in this way that this class in mathematics is going to be different. Plan very carefully for that first problem session. Choose a relatively simple problem and by all means make it one that is of obvious practical use. Save the more abstract problems for later in the course. Ensure that a high level of success will be attained with the first problem. Follow up the problem session in subsequent class meetings with some discussion of the content involved. These follow-up sessions can themselves illustrate, for example, a laboratory approach. Begin with perhaps one problem session a week (every third class meeting) and build up to two a week or more. Work from easy problems to more difficult, from practical to abstract, and begin to keep a file on good problem situations for the future. It may be that the college instructor will discover that the standard text can be covered with a combination of independent study, tutorial sessions with individuals or small groups, and a series of optional lectures where the students know the topics to be covered and the schedule of lectures. During a problem session the instructor, particularly at the beginning of the year, plays the role of a resource person, moving from team to team to offer suggestions when needed, to answer questions, or to take individual students aside for special help. As the year progresses, the instructor should find it less and less necessary to assist. Be very aware of the students who are good in mathematics and who may have had excellent training in high school. These students may be bored stiff by problems similar to the carpet problem. It is possible to identify these students very quickly. Make them assistants to help with hints, answer questions, or serve as tutors; or provide them with much more challenging problems that they can work on individually or with other bright students.

More on calculators

Calculators should be required in the content course for future elementary school teachers from the first day, and it is the instructor's responsibility to see that every student learns how and when to use a calculator. The instructor should also suggest to the future teachers how calculators can be used in the elementary school classroom with their own future students. Just as one cannot learn to be a problem solver without solving problems, it is improbable that future teachers will know how to use calculators properly in their teaching if they have not used them and been instructed in their use as students. The content and methods courses offer perhaps the only opportunity for the mathematics educator to influence the future teacher about calculators; consequently, they should be used regularly, especially in the content course. Find out at the beginning of the course how much resistance there is on the part of the future teacher to the use of calculators in the elementary school classroom. Then watch that resistance disappear as they discover how much more mathematics can be learned and taught, without sacrificing necessary skills, with the use of calculators. My own feeling is simple and strong—any college instructor not using calculators in preservice and in-service content and methods courses is doing a disservice to the future teacher.

The suggestions above have concentrated on the content course for elementary school teachers. What about the methods course? It is clear that teaching the content course in the manner suggested injects into it a great deal of methodology. That is a very large bonus. Where the methods and content courses have already been unified, these proposals will only strengthen the relationship. Where they are separate, the content course will have much more relevance to the methods course and vice versa—to everyone's advantage. The implication is that colleges will need a good deal more communication between the two sets of content and methods teachers, who are often in disjoint sets, to get the most mileage out of this approach.

What about in-service training? Clearly, in-service courses conducted at colleges or in school districts can be modified to fit the problem-solving mold. They should be; but the biggest opportunity, the place where the largest numbers of future teachers can be reached, is in preservice training.

REFERENCES

Barclay, Tim. "Buggy." *Mathematics Teaching,* no. 92 (September 1980), pp. 10–12.

Committee on the Undergraduate Program in Mathematics (CUPM). *Recommendations for the Training of Teachers of Mathematics.* Berkeley, Calif.: CUPM, 1961.

Gibney, Thomas, Len Pikaart, and Marilyn Suydam, eds. *Problem Solving—a Basic Mathematics Goal.* Columbus, Ohio: Ohio Department of Education, 1980.

Kemeny, John G., J. Laurie Snell, and Gerald L. Thompson. *Introduction to Finite Mathematics.* Englewood Cliffs, N.J.: Prentice-Hall, 1957.

Krulik, Stephen, and Jesse A. Rudnick. *Problem Solving: A Handbook for Teachers.* Rockleigh, N.J.: Allyn & Bacon, 1981.

Lappan, Glenda, and M. J. Winter. "Probability Simulation in Middle School." *Mathematics Teacher* 73 (September 1980): 446–49.

National Council of Supervisors of Mathematics. "Position Paper on Basic Skills." *Arithmetic Teacher* 25 (October 1977): 19–22.

National Council of Teachers of Mathematics. *Problem Solving in School Mathematics.* 1980 Yearbook. Reston, Va.: The Council, 1980.

Piaget, Jean, and Bärbel Inhelder. *The Origin of the Idea of Chance in Children.* New York: W. W. Norton, 1975.

Polya, George. *How to Solve It.* New York: Doubleday Publishing Co., Anchor Press, 1957.

Simon, Julian L., and Allen Holmes. "A New Way to Teach Probability Statistics." *Mathematics Teacher* 62 (April 1969): 283–88.

Recommendation 8

PUBLIC SUPPORT FOR MATHEMATICS INSTRUCTION MUST BE RAISED TO A LEVEL COMMENSURATE WITH THE IMPORTANCE OF MATHEMATICAL UNDERSTANDING TO INDIVIDUALS AND SOCIETY

26

A Model for Community and School Interaction

Helen Johnson King
Linda Pinson

BOTH educators and laypersons acknowledge the shortage of people with adequate mathematical training. More and more citizens are admitting handicaps in the area of mathematics as the public spotlight touches on such problems as math anxiety and math phobia, sexual and environmental bias, and the difficulty in motivating and training students in mathematics. In *An Agenda for Action: Recommendations for School Mathematics of the 1980s,* the National Council of Teachers of Mathematics recommends that public support for mathematics instruction be raised to a level commensurate with the importance of mathematical understanding to individuals and society.

To implement this recommendation, one high school mathematics department developed a model program that promotes constructive interaction among staff members, students, parents, and community members. The program is also designed to elicit public support for mathematics instruction and encourage the mathematically talented to fulfill their mathematical potential. The program has ten objectives, some to be implemented during the current school year and others to be developed over a longer period:

1. Develop mathematics career awareness and coordinate existing materials with new materials in the career center.
2. Establish a speakers bureau.
3. Establish a mathematics club and a local chapter of Mu Alpha Theta.
4. Develop a "math day" program.
5. Spotlight mathematics and mathematics programs in the school yearbook, school newspaper, school radio, and the local newspaper to spark interest in, and awareness of, mathematics.
6. Disseminate information on female participation in mathematics programs and careers.

7. Prepare a teacher's handbook.
8. Develop in-service programs on math anxiety, math phobia, female participation in mathematics, microcomputers, and recreational mathematics in the classroom.
9. Establish a microcomputer and software facility in the library.
10. Conduct a public clinic on math anxiety.

Development of Career Awareness

To implement the first goal, we used funds from a federal grant obtained by the school for career education. The mathematics department was surveyed to determine what materials for correlating career guidance and mathematics were needed and desired. Although career materials for consumer mathematics and general mathematics were readily available, materials for other levels were still needed. Information on sources was obtained from the National Council of Teachers of Mathematics, Society of Actuaries, Casualty Actuarial Society, Mathematical Association of America, Society for Industrial and Applied Mathematics, and the U.S. Government Printing Office. After lists of resources were collected and evaluated, materials were ordered. In addition, ideas for career resources were obtained from commercial educational catalogs.

When the materials arrived, they were displayed and reviewed at a mathematics department meeting so that all the teachers were aware of them. The teachers were then made responsible for the materials they thought most useful to them. A list of all materials and their assigned location was distributed to each member of the department.

Each mathematics teacher was encouraged to find a method for incorporating career education into the curriculum. As a result, career corners, newsletters, and bulletin boards were initiated. The following procedure proved very successful. First, the career guidance counselor spoke to the mathematics classes. Then, to individualize this learning experience, each student was assigned a brief period of class time to work in the career center finding information on two careers, one of his or her choice and one assigned by the teacher. The form in figure 26.1 helped to direct the activity.

All students discussed some aspect of their experience with the class. Reaction was very positive. Most students had not visited the career center prior to this assignment. Some students noted addresses on their forms for further information.

Speakers Bureau

The speakers bureau was developed to draw on community resources.

Name ______________________ Career Title __________

Job Description Earnings

Training (Education needed)

Working Conditions

Places of Employment

Helpful Subjects or Classes

Employment Outlook

Fig. 26.1

The form letter in figure 26.2 was sent to businesses listed with the local chamber of commerce.

Universities and colleges in the area were contacted, and their speakers bureaus furnished lists of relevant names. The career center gave additional names. Suggestions were sought from the National Council of Teachers of Mathematics. This list of possible speakers from the local community was compiled for the mathematics teachers. In this way, community interaction and public participation in mathematics education were encouraged.

Math Club

The first meeting of the Lakota High School Math Club was advertised by posters, teachers, and the school newspaper. Thirty-six interested students met to write a constitution, elect officers, and form a club whose purpose (according to its members) was to stimulate interest in mathematics, learn about mathematics and career opportunities, provide recreation and enjoyment of mathematics, and provide such services to the school as assistance with contests, a "math day," and scholarship information. From this nucleus

Dear Sirs:

The mathematics teachers of Lakota High School are interested in establishing a speakers bureau. We would like to draw on the resources of the surrounding community. We want to emphasize the importance of mathematics in industry and in the professions. The most effective manner to emphasize the importance of mathematics is for students to hear first hand what mathematics skills are used on the job and in various professions from people practicing those skills.

We would appreciate your assistance in establishing this speakers bureau. We feel that as future employers of our students who graduate from Lakota and from surrounding universities, you can play an important part in the motivation of these young adults.

If you have an existing speakers list available, please share it with us. If you can put together a list of possible speakers who would meet our interests, please send us the list. Thank you.

Sincerely,

Fig. 26.2

of highly motivated students a local chapter of Mu Alpha Theta was established. Appropriate publicity was released to the community and school newspapers, and photographs were included in the yearbook.

Math Day

Math Day is a normal school day offering concentrated activities in mathematics. The purpose of a math day is to promote interest in, and awareness of, mathematics, to involve parents and community members in the mathematics curriculum, and to motivate students in mathematics through mathematical recreations, career information, community speakers, and mathematics contests. The day's activities were publicized by sending letters of invitation to the chamber of commerce, local ministers, members of the board of education, local junior high schools, other area high schools, and county officials. Articles were submitted for publication in the local newspapers (fig. 26.3) and the school newspaper. Signs made by club members were prominently placed in the community and in the school. The whole faculty was kept aware of Math Day plans so that they could also participate.

Math's No. 1 at Lakota Friday

Math Day will be observed Friday, April 10, at Lakota High School.

Speakers from engineering, insurance, local business, and MU's math speakers branch will give presentations at the high school in the day-long event which begins at 8:30 with a coffee in the media center.

Topics include recreation, careers and employment, enrichment, and consumer education.

The media center welcome for guests and parents will be held between 8:30 and 9:30 a.m.

Faculty members will speak on scholarships available in math and related fields, careers using and requiring math, and supplementary materials available in the media center.

MATH DAY ACTIVITIES

7:30 to 8:15 a.m.—coffee in the high school media center
7:30 to 10:30—math students field day competition
9 to 10 a.m.—welcome and faculty presentation
9:40 to 10:30—tours and in-school assemblies
10:30 to 2:15—guest speaker presentations

At the day's end, winners of the field day competition will be announced and prizes awarded.

SPEAKERS

Professor Robert Dieffenbach, MU—*Mathematical Games*
Professor Joe Kennedy, MU—*Tessellations*
Professor Richard Laatsch, MU—*Euclid Was a City Boy*
Jack Flaherty, Senour-Flaherty Insurance—*Math in Insurance*
Dan McCluskey, McCluskey Chevrolet—*So You Want to Buy a Car*
Carol Wright, General Telephone—*Math in Marketing*
Kelly Woodruff, Cincinnati Milacron—*Linear Systems and Programming for Robot Commands*

Proctor and Gamble representative:
Calculus and Higher Math in Business Research, Computer Technology, and Data Processing

Fig. 26.3

The day's activities began for parents and community members with coffee and cake provided by the parent, teacher, and student association at 7:30 A.M. in the school media center. A welcome to those attending and a mathematics curriculum overview was followed by an orientation with the career-center counselor. A representative of the guidance department provided computer printout sheets itemizing financial assistance available for students interested in pursuing an education in mathematics. The librarian discussed the audiovisual aids, books, pamphlets, and magazines about mathematics that the school library provides for teacher and student use. This information was followed by tours throughout the building. The tours were conducted by members of the mathematics club. Parents were invited to stay for the afternoon sessions and hear community speakers provided by Milacron, General Motors, Bell Telephone, and Miami University.

While the open house was in progress, a mathematics competition for individual students was held in the cafeteria. This contest was divided into the following categories: *(a)* general mathematics, consumer mathematics, basic algebra, and basic geometry; *(b)* algebra 1; *(c)* geometry; *(d)* algebra 2; and *(e)* advanced mathematics. Students participating in the contest signed up for the highest group for which they were eligible, and calculators were permitted. The total competition consisted of two sections: group activity of puzzles and team problems and individual tests. The group activities were conducted the day before Math Day and the individual tests on the morning of Math Day. Members of the first-place groups in all five categories won gift certificates. Individual winners received trophies. Club members assisted mathematics teachers in proctoring the tests and grading the results so that scores and winners could be announced the same day.

In addition to the mathematics contest in the morning, there were two slide presentations on careers in mathematics and a speaker on computers. Additional speakers were available in classrooms during the last three periods of the day. Their presentations dealt with such topics as the application of mathematics to marketing, linear systems and programming for robot command, purchasing a car, computer applications in business, games and recreations, tessellations, and one entitled "Euclid Was a City Boy."

Future Objectives

Initial steps have been taken for the preparation of a teacher's handbook, an in-service program for teachers on computers, and funding for a microcomputer facility. The continued constructive interaction of teachers, parents, students, and community members is needed for the accomplishment of these goals.

Constructive interaction encourages positive involvement, and positive involvement encourages public support. Thus, instead of saying, "That would never work in our community," the community will say, "How can we make it work?"

27

Back to the Real Basics

Ann Kahn

IT would be difficult to quarrel with the proposition that public support for mathematics instruction should be raised to a level commensurate with the importance of mathematical understanding to individuals and society. But that proposition raises profound questions. What is the level of importance of mathematics learning and, indeed, of public education in general? What, specifically, constitutes the level of support we might agree on? What does public support entail besides the obvious financial support?

We all have our own special agendas and interests, but those of us who view education from the perspective of parents and school board members share larger goals with you who are professional educators. We want the children and youth of this country—for whom we are responsible individually as parents and teachers and collectively as society—to learn those concepts and skills that give them the greatest opportunity to realize their potential as productive and fulfilled human beings and good citizens. Certainly we recognize that the concepts and skills of mathematics make an essential contribution to that learning.

However, the fact that education is important for everyone and is one of society's tools for maintaining itself does not ensure a smooth path for the school as a public institution. Schools today are caught in a series of conflicting objectives, the pressures of inflation, eroding budgets, and a rising tide of expectations from both parents and community about what education should be accomplishing. There are even serious threats to the concept of public school itself.

The ideal of the public school is a remarkable contribution of American democratic thought. It recognizes that society has the obligation to educate its youth. Education, then, is not solely a responsibility of individual parents toward their children. Everyone has a stake in schooling future generations and thus an obligation as a societal whole. The phrases *public support* and *importance to society* underscore the currency of the ideal. They remind us of the crucial role of a healthy system of public education in a free society.

Curiously enough, threats to public education are coming at a time when the public school, for the first time in history, is about to reach its historic goal of universal education. In 1920, for example, only 20 percent of our youngsters in their teens went on to high school. In 1950, that number reached 75 percent, and in 1980 we saw over 90 percent of the teenagers in this country going on to high school. The United States now leads the world in the percent of children who are enrolled in school; Thomas Jefferson's concept of a democratic society that is based on an educated electorate may really be coming to pass.

The absorption of these additional students, who in previous years left school long before high school, is bringing some real opportunities and some critical problems. We are no longer dealing with the top 20 percent, as high schools in the 1920s did. We are not dealing with the cream of the crop and accepting the other 80 percent as a pool of unskilled laborers. The world that young people now enter has very different needs, and even those needs are changing so rapidly that it is hard for school systems to avoid training people for obsolescence.

The demands on public education are enormous, and they have caused severe repercussions. The willingness of the public to provide the financial support necessary to a viable institution of public education has been questioned. This question surfaced in the proliferation of tax restrictions beginning with the much-publicized Proposition 13 in California. In Virginia the restriction is known as "TEL," a tax exemption limitation. By whatever name they are known, the pressures in many states and localities to place funding lids on either taxation or the use of local spending have serious impact on the reasonable funding of public schools. Many of these proposals, as initially passed, were buffered by state surpluses that could take up the slack. As the country moved into a recession and state surpluses disappeared, the effect on local public school systems has sometimes been severe.

The combined effect of such actions as Proposition 13 could seriously dilute public school services. And the proposals, attractive as they may sound, have the potential to destroy the ability of the public school systems to meet even their minimum needs, thereby forcing parents, if they can afford it, into choosing private rather than public schools. Those who cannot afford it will have to settle for a less satisfactory education as diminished funding provides less than adequate staffing.

Ironically, the very success of the public schools in establishing virtually universal education contributes indirectly to their problem of declining public support. Expectations for results remain the same, whether the school is educating over 90 percent of the teenage population or only the top 20 percent. In fact, these expectations are higher now, when schools are asked to deal with much more than just academic learning. With often unrealistic demands, it was inevitable that there would be public disenchantment.

The movement away from public education has been reflected in proposed initiatives that challenge the very concept of the public school. One such idea is the federal tuition tax credit, which partially subsidizes attendance at private schools. This legislative concept substantially changes the political balance between public and private education and alters the nature of a society's compact to educate its future generations.

The notion that public and private education can compete equally under tuition tax credits does not stand up to scrutiny when the issue of selective admissions is raised. Private schools can be selective; public schools cannot. It is legitimate to ask whether such a plan will inevitably result in a public school filled with only the poorest and hardest-to-educate children. Today, these profound philosophical questions remain unresolved and too often ignored.

A specific recommended action included in the *Agenda* under the broad category of public support says:

> Parents, teachers, and school administrations must establish new and higher standards of cooperation and teamwork toward the common goal of educating each student to his or her highest potential.

Some feel that school problems cannot be resolved jointly by parents and teachers. Indeed, some academicians feel that the views of parents and teachers are so divergent that there is very little on which we can agree. Many parents are turned off by the present state of education and seek to assess blame, holding the school system responsible for what they see as serious failures. Why is there this terrible frustration with education? How deep is the public loss of confidence in education?

What is the public's conception of an ideal school? A profile of such a school was developed by George Gallup in a 1979 poll. The parents' first concern was that teachers be well qualified, that they pass state board examinations before being hired and at regular intervals thereafter, and that they take a personal interest in each of their students and motivate them to succeed. Parents expect discipline to be strict and based on specifics agreed on by parents and teachers. The curriculum must emphasize the basics but include a very broad definition of what the basics are; vocational training should be offered for those who do not go on to college.

Schools should communicate more with parents and the community, and teachers should confer with parents about specific children. Courses or seminars should be organized to help parents learn how to help their children in school and deal with issues like drugs and alcohol, emotional problems, and serious behavioral problems. All this is a big order, but public concerns must be heeded if we expect public support.

It is interesting how directly the *Agenda* addresses these public concerns.

Consider the following excerpts from that document and compare them with the list of concerns cited above.

- School boards and school administrations should take all possible means to assure that mathematics programs are staffed by qualified, competent teachers who remain current in their field.
- Teachers must be sensitive to the needs of their students and dedicate themselves to the improvement of student learning as their primary professional objective.
- Parents must support the maintenance of agreed on standards of achievement and discipline.
- There must be an acceptance of the full spectrum of basic skills and recognition that there is a wide variety of such skills beyond the mere computational.

If is often said that a significant cause of the erosion in public confidence in its schools is declining test scores, usually standardized, norm-referenced test results. For years the public accepted these tests uncritically as the sole measure of the success or failure of the school. Parents, myself included, accepted administrators' views that although there might be some flaws in the testing process and in our heavy reliance on testing, we had to go along with it because we really had no alternative. Many parents are no longer willing to accept that evaluation. We are aware now, as more light is shed on the development and use of tests, that there is a significant reduction in confidence that these tests either serve as reliable indicators of success or give us the timely diagnostic information that we should have to help children progress.

As parents, we are concerned that many of the current testing instruments give us an inadequate reflection of what a child knows and do not assist teachers to respond appropriately. We are concerned that tests are designed in a multiple-choice format, not because that is the best way to learn what children know, but because it is the easiest way to score and administratively handle the information. This causes serious problems because it limits the range of what is being tested. In effect, it penalizes some children who are developing the attitudes that we strive for in the schools—creativity, curiosity, and divergent thinking—the very qualities that exist at the higher realms of every discipline in the school system. Many tests reward instead the lower levels of rote memorization.

As parents, we are concerned that the power to test is the power to alter the curriculum. We would like the reverse to be true: tests should follow the decisions of educators; the curriculum should not follow the test.

As parents, we question whether the dominance of norm-referenced standardized testing is really responsive to the major goal of the public school system, which is the improvement of the education of children, or whether it simply satisfies educators by sorting children into more manageable groups. We are concerned that more and more classroom time is being spent in testing at an enormous cost in both educational dollars and instructional time. Most important of all, the present system of testing worries us because we know that children are not always tested on what they have been taught and that when they are it is sometimes a matter of directing all instruction toward a narrow range of test items.

This tyranny of testing has resulted in a concentration on narrow skills in the name of "back to the basics." Schools are limiting their emphasis in mathematics to computational skills. As computation scores have improved, there has been a decline in the ability of students to apply those skills in any practical sense. Roy Forbes, director of the National Assessment of Educational Progress, has stated that "schools are not doing a very good job training students to analyze and solve problems." He also said, "NAEP data parallel the SAT results in that students do not have the analytical or comprehension skills they did 10 years ago."

As parents, many of us would agree with another of the *Agenda* recommendations that "the success of mathematics programs and student learning must be evaluated by a wider range of measures than conventional testing" and would extend the statement to other disciplines as well. We know that other alternatives, some more responsive to the needs of teachers and to the broader evaluation of children's progress as well, are beginning to be made available. We look to mathematics educators to help develop alternative methods of assessment that will more fairly reflect a child's true knowledge.

It is interesting to review how different elements of the public think the schools are measuring up. A Gallup poll reports that young people, for example, rate the public schools much higher than adults do. Students and young people still consider America the land of opportunity, and this is true regardless of their race, social background, or geographical area. They are increasingly work oriented, but they are concerned about salary rather than prestigious occupations. Only about 10 percent of children now indicate that they enjoy school very little. Young people feel that students are not required to work hard enough in school, and they overwhelmingly think that damage done by juvenile vandals, for example, should be paid for by the vandals themselves and not by citizens at large. In short, a more responsible attitude about school and about life seems to be emerging in the public schools, where we have the most representative slice of American life of any institution in this country.

Parents, too, are beginning to appraise much more realistically what schools can and should be expected to accomplish, and they continue to have a great desire to participate in that process. John Gardner some years ago warned about the dangers in a democratic system of being caught between what he called "unloving critics and uncritical lovers." Pick up any recent newspaper or news magazine and you will find many of education's unloving critics. But parents do not intend to play the role of uncritical lovers either. That is too high a price to pay for participation in the system. We, teachers and parents, have a role to play together, and like most healthy relationships it will not suffer from either honesty or civility when we deal with our differences.

These problems are not specific to mathematics, but they are relevant to achieving the goals for mathematics education set out in the *Agenda for Action.* Certainly mathematics learning is basic and essential to everyone. A Gallup poll indicated that 97 percent of the public agrees. There is no doubt that mathematical understanding and skill are increasingly in demand in our society. Considerable attention has been given of late to a growing crisis in mathematics and science education and its relationship to national priorities in a technology-dominated world.

We have also been made aware of the critical shortage of qualified mathematics teachers. These are all public problems and relate to public support. Mathematics teachers are part of this public and must be involved not only in the unique concerns of mathematics education but in the larger context in which these concerns about education reside. They have a larger responsibility as professional educators and, more importantly, as people who have chosen a career that means they care what happens to kids. More than a job is at stake when an entire system of education is threatened. The foundation of this nation relies on education as the cement that holds us all together.

During the celebration of our bicentennial there was a popular sign that read "America ain't perfect but we ain't finished yet." I believe that, just as I believe the conclusion of a prayer that states, "Lord, we know we ain't what we should be, and, Lord, we know we ain't what we're going to be, and, Lord, we know we ain't what you want us to be, but remember, Lord, we ain't what we used to be either." This country has come a long way in 200 years, and education is largely responsible for our miraculous progress.

If this nation has a true commitment to a pluralistic society, then public education is going to be a barometer by which we test that pledge. But that commitment cannot be realized if either parents or teachers in any discipline stay aloof. Clearly, the real basics are not reading and arithmetic. The real basic, which calls for the united concern and joint efforts of teachers and parents, is to guard this unique system that makes education available to everyone so all can function effectively in ensuring a free democratic society. It requires our joint awareness that we *all* must be involved in these critical educational issues.

NCTM's *Agenda for Action* recognizes this need for a coordinated effort, the cooperation of public and professional:

> The professional community and society share a common goal: to bring all

> citizens to the full realization of their mathematical capacity. This . . . requires the commitment and cooperation of all segments of society.

I should simply suggest that the statement be broadened to include the full realization of all our capacities, to education itself. The *Agenda* provides the framework for an organization of teachers to be a part of the response to all the issues I have raised here. It is my fervent hope that you will use that document as the basis for an active, cooperative effort with educators of other disciplines, with boards of education, and with parents to achieve these worthwhile goals in this decade.